This work: "Silicon-organic hybrid devices for high-speed electro-optic signal processing", was originally published by KIT Scientific Publishing and is being sold pursuant to a Creative Commons license permitting commercial use. All rights not granted by the work's license are retained by the author or authors.

Cover Image Credit/Copyright Attribution: IIIerlok_Xolms/Shutterstock

Silicon-organic hybrid devices for high-speed electro-optic signal processing

by
Matthias Lauermann

Karlsruher Institut für Technologie
Institut für Photonik und Quantenelektronik

Silicon-organic hybrid devices for high-speed
electro-optic signal processing

Zur Erlangung des akademischen Grades eines Doktor-Ingenieurs
von der KIT-Fakultät für Elektrotechnik und Informationstechnik des
Karlsruher Instituts für Technologie (KIT) genehmigte Dissertation

von Dipl.-Ing. Matthias Lauermann geboren in Schwäbisch Hall.

Tag der mündlichen Prüfung: 9. März 2017
Hauptreferent: Prof. Dr.-Ing. Christian Koos
Korreferenten: Prof. Dr.-Ing. Dr. h. c. Wolfgang Freude
　　　　　　　　Prof. Dr. rer. nat. Uli Lemmer

Karlsruher Institut für Technologie (KIT)
KIT Scientific Publishing
Straße am Forum 2
D-76131 Karlsruhe

*This document – excluding the cover, pictures and graphs – is licensed
under a Creative Commons Attribution-Share Alike 4.0 International License
(CC BY-SA 4.0): https://creativecommons.org/licenses/by-sa/4.0/deed.en*

Table of contents

Kurzfassung v

Preface ix

Achievements of the present work xiii

1 Introduction 1
    1.1 Silicon photonics in optical communication links with high capacity 2
    1.2 Silicon photonic frequency shifter 5

2 Theoretical background 7
    2.1 Second-order nonlinearities and the electro-optic effect 7
        2.1.1 Electric polarization 7
        2.1.2 Nonlinear polarization and Pockels effect 9
    2.2 Electro-optic modulators for advanced modulation formats 11
        2.2.1 Mach-Zehnder modulator 11
        2.2.2 In-phase – quadrature modulator 13
        2.2.3 Advanced modulation formats 14

3 Energy efficient high-speed IQ modulators 17
    3.1 Silicon-organic hybrid integration 17
        3.1.1 Concept of resistively coupled SOH modulators 17
        3.1.2 Capacitively coupled SOH modulators 24
        3.1.3 Bandwidth of SOH modulators 27
        3.1.4 Organic electro-optic materials 33
    3.2 State of the art silicon based IQ modulators 36
        3.2.1 IQ modulators based on the plasma dispersion effect 36
        3.2.2 IQ modulators based on SOH integration 38

3.3 Fabrication of SOH IQ modulators 39
3.4 Generation of advanced modulation formats
at low power consumption 40
3.5 Generation of data signals with high symbol rates 53

4 Frequency shifter integrated on the silicon platform 65
4.1 Frequency shifting in waveguide based modulators 65
4.1.1 Single-sideband modulation for frequency shifting 66
4.1.2 Single-sideband modulation with phase modulators 67
4.1.3 Small signal modulation and temporal shaping for
enhanced side-mode suppression ratio 69
4.1.4 Frequency shifting via serrodyne modulation 72
4.2 State of the art 73
4.3 Silicon-organic hybrid frequency shifter 75

5 System implementation of SOH devices 95
5.1 Operation at elevated temperatures 95
5.1.1 Static characterization of SOH devices with SEO100 96
5.1.2 Signal generation with silicon-organic hybrid
modulators at increased temperature 98
5.2 Fabrication of SOH devices on an
industrial silicon-photonic platform 108
5.2.1 Special requirements of SOH devices and their
implementation in a commercial process flow 109
5.2.2 Characterization of the fabricated devices 111

6 Summary and outlook 115
6.1 Summary 115
6.2 Outlook and future work 117

A Mathematical relations 120
A.1 Fourier transformation 120
A.2 Hilbert transformation 120

B   Electrical transmission lines 121

    B.1   Lumped element circuit model of the electrical transmission line ... 121

    B.2   Impedance matching and reflections 123

    B.3   Lossy transmission lines 123

C   Electro-optic bandwidth measurement 125

D   Field calculations for single-sideband modulation 126

    D.1   Single-sideband modulation without temporal shaping 126

    D.2   Single-sideband modulation with additional harmonics 127

E   Bibliography 129

F   Glossary 144

    F.1   List of abbreviations 144

    F.2   List of mathematical symbols 146

          Greek symbols 146

          Latin symbols 147

Danksagungen 151

List of publications 155

    Patents 155

    Book chapters 155

    Journal publications 155

    Conference publications 158

Curriculum vitae 165

Kurzfassung

Die vorliegende Arbeit befasst sich mit elektrooptischen Modulatoren auf der Basis von integrierten optischen Schalkreisen auf Silizium. Dabei steht vor allem die Manipulation des komplexwertigen optischen Feldes zur Signalprozessierung mit höchsten Geschwindigkeiten, bei gleichzeitiger Reduktion des Energieverbrauchs im Vordergrund. Die Silizium-Organik Hybridintegration (SOH) wird hierbei genutzt um die Eigenschaften passiver Siliziumwellenleiter mit einem elektrooptisch aktiven organischen Material zu ergänzen. Mit Hilfe der SOH-Integration werden in dieser Arbeit Modulatoren in Inphasen-Quadratur (IQ) Konfiguration hergestellt und charakterisiert. Die Erzeugung von Datensignalen mit Datenraten über 100 Gbit/s wird demonstriert, bei gleichzeitiger Reduktion des Energieverbrauchs auf nie zuvor erreichte Werte. Nicht nur in der Telekommunikation, auch in anderen Feldern kann die hohe Bandbreite und der geringe Energieverbrauch von SOH-IQ-Modulatoren genutzt werden. Dazu wurde ein integriert-optischer Frequenzschieber auf SOH-Basis entwickelt, der unter anderem in der Messtechnik verwendet werden kann. Ein weiterer Aspekt dieser Arbeit ist die Anpassung der SOH Modulatoren auf die Erfordernisse der Systemintegration. Das Design der Bauteile wird dazu an die Standards kommerziell verfügbarer Fabrikationsprozesse angepasst sowie die Funktion der SOH Modulatoren bei erhöhten Temperaturen untersucht.

Die Integration optischer Schaltkreise, verknüpft mit der integrierten Elektronik, bildet die Grundlage für zukünftige Systeme und neue Anwendungen in vielen unterschiedlichen Bereichen. Während die integrierte Elektronik in den letzten Jahrzehnten eine enorme Entwicklung durchlaufen hat, stehen integrierte optische Schaltungen vergleichsweise noch am Anfang ihrer Entwicklung. Es ist jedoch absehbar, dass auch in der Optik integrierte Schaltungen die Systeme aus diskreten Bauteilen ablösen werden. Vor allem die Telekommunikationsbranche ist hierbei eine Triebfeder für die Entwicklung: Datenverbindungen mit immer höheren Bandbreiten erfordern immer schnellere und energiesparendere Systeme, bei gleichzeitiger Reduktion der Größe.

Kurzfassung

Silizium als Fabrikationsplattform ermöglicht eine einfache Realisierung integrierter optischer Schaltungen. Neben guten wellenführenden Eigenschaften des Materials kann hier auch auf die standardisierten CMOS Fabrikationsprozesse der Elektronik zurückgegriffen werden. Analog zu der integrierten Elektronik sind schon erste kommerzielle Foundries entstanden, die eine Fertigung von integrierten optischen Schaltungen auf Silizium nach Kundenwunsch ermöglichen. Hierbei kann auf Designbibliotheken zurückgegriffen werden, so dass grundlegende Funktionen wie verlustarme passive Wellenleiter, Photodetektoren auf Basis von Germanium oder auch elektrooptische Modulatoren direkt verfügbar sind. Jedoch können noch nicht alle Elemente auf der Siliziumplattform sämtliche Anforderungen optimal erfüllen. Vor allem der elektrooptische Modulator, typischerweise durch einen *pn*-Übergang im Wellenleiter realisiert, hat nicht die gewünschte Effizienz welche für eine hohe Integrationsdichte benötigt wird. Auch die gleichzeitige Kontrolle der Amplitude und Phase des optischen Signals ist mit reinen Siliziummodulatoren nicht optimal möglich.

Im Hinblick auf diese Probleme werden in der vorliegenden Arbeit IQ-Modulatoren basierend auf der SOH Integration entwickelt und untersucht, die verbesserte Eigenschaften bezüglich Geschwindigkeit, Energieverbrauch und gleichzeitige Ansteuerung der Amplitude und Phase bieten. Um eine breite Nutzung der SOH Technologie zu ermöglichen, wird die Ausführung der Modulatoren an die Anforderungen standardisierter Foundry-Prozesse angepasst.

Kapitel 1 gibt eine Einführung in integrierte optische Schaltungen, mit dem Schwerpunkt auf siliziumbasierter Photonik für die Nutzung in der Telekommunikation und als Frequenzschieber.

In *Kapitel 2* werden die theoretischen Grundlagen integrierter elektrooptischer Modulatoren erläutert. In kompakter Form werden nichtlineare Effekte zweiter Ordnung sowie interferometrische Strukturen zur Modulation des komplexwertigen Feldes diskutiert.

Kapitel 3 befasst sich mit SOH-IQ-Modulatoren, welche die Erzeugung von Datensignalen mit höchsten Datenraten bei gleichzeitiger Reduktion des Energieverbrauchs ermöglichen. Das Konzept der SOH-Integration wird im Detail erläutert, der Stand der Technik wird diskutiert und neuartige Konzepte für SOH Modulatoren mit erhöhter Bandbreite werden vorgestellt. Mithilfe von

SOH-IQ-Modulatoren wird bei der Erzeugung von komplexen Datensignalen ein zuvor unerreichter niedriger Energieverbrauch von 19 fJ/bit realisiert.

In *Kapitel 4* wird ein Frequenzschieber vorgestellt, welcher ebenfalls auf der SOH Technologie basiert. Die Grundlagen der Einseitenbandmodulation als Funktionsprinzip des Frequenzschiebers werden zu Beginn diskutiert und ein neues Konzept zur Verbesserung der Signalqualität wird untersucht. Im Experiment wird der integrierte SOH Frequenzschieber charakterisiert und die Verschiebung eines optischen Trägersignals um bis zu 10 GHz demonstriert.

Kapitel 5 umfasst unterschiedliche Aspekte, welche für eine breite Nutzung von SOH Bauelementen und für eine Implementation in integrierte Systeme notwendig sind. Das Design der SOH Bauelemente wird hierzu an die Anforderungen einer kommerziellen Foundry angepasst und die gefertigten Bauteile werden evaluiert. Um eine größere Stabilität bei höheren Temperaturen zu erreichen werden in diesem Kapitel organische elektrooptische Materialien im Hinblick auf ihre Temperaturstabilität untersucht. Mithilfe eines solchen Materials wird ein SOH Modulator realisiert, der auch bei 80 °C Umgebungstemperatur die Erzeugung von Datensignalen mit höchster Geschwindigkeit ermöglicht.

In *Kapitel 6* werden die Ergebnisse dieser Arbeit zusammengefasst und die zukünftigen Möglichkeiten und Entwicklungen der SOH Technologie umrissen.

Preface

Photonic integration conjoined with large scale electrical integration is a key technology for future communication and other technological systems. The integration of optical devices can not only play the role of supplementing the electrical circuitry with added functionality, it can in fact enable the creation of whole new types of systems and applications. While the field of optical communication with its demand for an ever increasing data rate is already heavily relying on optical technology and pushing towards photonic integration, also other fields such as metrology, sensing or medical applications are starting to leverage the advantages of optical integration [1].

Silicon photonics is considered to be a very well-suited candidate for such large scale optical integration. It allows the realization of photonic integrated circuits (PIC) with a high density, and the fabrication can rely on well-developed CMOS processes. Furthermore, an easily accessible infrastructure for fabrication has already emerged: commercially available foundry services together with process design kits (PDK) allow the easy realization of custom silicon PICs at low cost. While a large variety of optical devices such as low-loss passives, Germanium photodiodes or electro-optic modulators are available on the silicon photonic platform, some challenges still remain on the path towards high-performance and high-density integration.

The electro-optic modulator is a key device for PICs since it provides a direct interface between the electronics and the optical circuit. Furthermore, today's systems do not need just a simple on-off control for the light, but rely on the modulation of both, amplitude and phase of the optical field. Silicon itself does not provide any appreciable second-order nonlinearity due to its crystal symmetry. Modulators based on the plasma dispersion effect are therefore commonly used to achieve phase modulation in silicon waveguides. Using the plasma dispersion effect for phase modulation leads, however, to undesired amplitude-phase coupling of the optical signal and to a relatively low modulation efficiency. A low modulation efficiency is detrimental to high integration density: it requires a larger footprint of the device and larger drive voltages which in turn leads to higher energy consumption. The requirements for the next generation of silicon-based advanced modulators can therefore be summarized as follows: low energy consumption, high bandwidth, complex field modulation, and the ability to be fabricated in a standard silicon photonics foundry process.

The work in this thesis targets those issues of energy consumption, high bandwidth and compatibility with commercial processes. Using silicon-organic hybrid (SOH) integration, electro-optic modulators for complex signal processing are developed on the silicon photonic platform. They provide record-low energy consumption and can be used for optical communication at highest data rates as well as for other applications such as frequency shifting. To enable the implementation of SOH modulators in more complex systems, the device design is adapted for commercial silicon-photonics foundry processes, and organic materials are investigated which allow operation of such devices at elevated temperatures. This thesis is structured as follows:

Chapter 1 gives an introduction in optical integration, especially silicon photonic integration for optical communication links and frequency shifter.

Chapter 2 reviews the theoretical background of integrated modulators based on second-order nonlinearities. The second-order nonlinear effect is discussed, together with interferometric structures for advanced modulation.

Chapter 3 introduces SOH IQ modulators capable of generating complex signals at highest speed and lowest power consumption. The SOH concept is discussed, state-of-the-art devices as well as novel device concepts for highest performance. Using SOH IQ modulators, the generation of advanced modulation formats is demonstrated leading to a record-low energy consumption down to 19 fJ/bit.

Chapter 4 presents a frequency shifter realized with SOH integration. The theory of single-sideband modulation for frequency shifting is discussed, and a new concept to improve the signal quality by employing shaping of the drive signal is introduced. In an experimental demonstration, frequency shifts up to 10 GHz are realized.

Chapter 5 covers different aspects needed for a system implementation of SOH-based devices. The SOH design is adapted for fabrication in a commercial silicon-photonic foundry and the resulting devices were characterized. Furthermore, electro-optic materials are investigated to achieve a higher environmental stability of the SOH devices. Experimentally, the operation of an SOH modulator at 80 °C was demonstrated while generating high-speed data signals.

Chapter 6 gives a summary of the work in this thesis and outlines further steps in the development of SOH-based PICs.

Achievements of the present work

In this thesis integrated electro-optic modulator structures were investigated with the goal to realize high-speed processing of the complex optical field while maintaining a low energy consumption. Silicon-organic hybrid (SOH) integration is used to demonstrate high speed IQ modulators with lowest energy consumption as well as application as a broadband frequency shifter.

A concise overview of the main achievements is given in the following:

Advanced modulation with low power consumption: An SOH-based IQ modulator was demonstrated with a record-low energy consumption of 19 fJ/bit for 28 GBd 16QAM modulation [J7], [C16], [C30].

High speed signal generation: While maintaining a low energy consumption, 16QAM modulation was realized with symbol rate of 40 GBd, and 4ASK modulation with a symbol rate of 64 GBd [J9], [J16], [C21], [C28], [C30].

Organic materials for SOH integration: SEO100 was introduced in the SOH platform, providing higher temperature stability. The poling process was optimized to achieve high efficiency in SOH devices [J16], [C28].

Operation of SOH devices at elevated temperatures: An SOH modulator was demonstrated, capable of generating advanced modulation formats at highest speed at an elevated temperature of 80 °C [J16], [C28].

Capacitive coupled SOH: To improve the performance of SOH devices, a new concept of capacitively coupled SOH modulators was developed. By adding materials with high permittivity to the SOH platform, higher modulator bandwidth can be achieved.

SOH devices on a commercial foundry platform: The design of SOH modulators was transferred to meet the standards of a commercially available 248 nm silicon photonic platform. High performance SOH devices were realized using such a process [C34], [C37].

Design of silicon photonic PICs: Mask layouts for various photonic devices and applications were designed and aggregated on densely integrated silicon photonic PICs, fabricated by foundry services [C17], [C18], [C41].

Achievements of the present work

Silicon photonic frequency shifter: An integrated frequency shifter was demonstrated using SOH integration, enabling frequency shifts of up to 10 GHz [J17], [C14].

Optimized frequency shifting: Introducing temporal shaping of the drive signal for single-sideband frequency shifter, a high conversion efficiency was achieved while maintaining a high SMSR [J17], [C19].

1 Introduction

The past century is often named as "century of the electron", dominated by the outstanding development of electronics and microelectronics with its tremendous influence on our daily life. The current century, however, is often dubbed as "century of the photon", and it is expected that photonics will make similar transforming changes to our society as electronics did in the past. And indeed, many fields are already being transformed by modern photonic technologies such as sensing, metrology, and perhaps most dominantly telecommunication. Key for the development of electronics was the step from individual components such as single vacuum tubes or single transistors to integrated electronic circuits with nowadays billions of transistors merged on a single chip [2]. The same process has already started in photonics: Photonic integration is the idea to combine multiple optical elements on one monolithic photonic integrated circuit (PIC), and thereby leveraging the advantages of size reduction, cost reduction and monolithic fabrication to realize more complex optical circuits in smaller devices. In photonics, an exponential increase in integration density can already be observed [3], similar to the one described by Moore's law in electronics.

Different material systems are used to create photonic integrated circuits, and each platform has its specific advantages and disadvantages. The silica-on-silicon waveguide platform offers low-loss waveguides and easy fabrication, but due to its low index contrast it has a rather large footprint, and the material systems do not offer any appreciable second-order nonlinearity for phase modulation, nor does it allow the generation or detection of light. Lithium niobate ($LiNbO_3$) has a moderate second-order nonlinearity but has also a large footprint and offers no active components. Indium phosphide (InP) is maybe the most complete photonic platform, offering active devices such as lasers and photo detectors, absorption and phase modulators, as well as passive elements [4]. However, the main drawback is the complicated fabrication. The material itself is scarce and expensive, and the wafers have to be grown in a complicated epitaxial process. Together with the relatively small wafer size between two and four inch, it is difficult to transition to high-volume mass fabrication [5]. Silicon-on-insulator (SOI) as photonic platform, however, can make use of all the standard processes developed for CMOS electronics. The high index contrast of the SOI platform to allows for very compact devices, and together with large wafer sizes of 12 inch and above, high volume manufacturing is a given. Silicon does not offer a direct bandgap, which would be necessary to realize a laser, however, Germanium can easily be integrated in a standard CMOS process to fabricate a detector [6]. Phase modulators are typically realized by using the plasma dispersion effect [7]. One major advantage of silicon photonics is the availability of foundry services. By offering a qualified fabrication process and a whole library of standard elements, such foundry services offer an easy entry point into photonic integration and make it widely available [8]. Silicon photonics is therefore a promising platform for a large variety of PICs spanning various applications.

One key element for PICs is the electro-optic (EO) modulator, it provides the direct interface between the electrical circuits and the optical circuits. Among those, phase modulators are the most versatile devices. When embedded into an optical interferometer they allow to control the optical field with electronic signals. This is important for realizing complex functionalities on the PIC, such as advanced optical signal processing. As research is focusing on PICs with higher speed and lower energy consumption, this directly relates to the requirements for EO modulators: Highly efficient modulators are paramount for a low energy consumption as well as for compact, densely integrated devices. Furthermore, to keep up with the increasing speed in electronics, the speed of the electro-optic modulators has to increase accordingly.

1.1 Silicon photonics in optical communication links with high capacity

The field of optical communication is driven by the steadily increasing demand for high-capacity data transmission, mainly generated by today's cloud-based services. Large scale data centers serve hereby as central nodes of the cloud. As reported in the Cisco global cloud index 2015 – 2020 [9], the data center IP traffic will see a threefold increase from 4.7 ZB up to 15.3 ZB from 2015 to 2020.

It is important to note that about three quarters of this traffic will stay within the data centers. To scale the network infrastructure with this increase in traffic, the capacity of each link has to be increased, especially within data centers, where the physical fiber connections have already become a scarce resource and have to be used as efficiently as possible.

Coherent transmission together with advanced modulation formats are a good approach to increase the bandwidth of an optical link [10]. Due to its more complex implementation of coherent transmission, it was first deployed in long-haul communication using discrete optical components, but it has become more and more attractive also for medium and short-reach links [5]. Optical integration and especially silicon photonics with its dense integration offers a great opportunity to handle the complexity of the implementation on a PIC. Electro-optic in-phase/quadrature phase (IQ) modulators are the key element to generate signals with advanced modulation formats and can be realized on the silicon photonic platform with a small footprint. The CMOS based fabrication technology enables high-volume fabrication at low costs.

Several implementations of IQ modulators have been already demonstrated on the silicon photonic platform, relying mainly on modulating the density of free carriers in an silicon waveguide to achieve an optical phase modulation [11-13]. While the performance of those modulators is good, they suffer from the low efficiencies, resulting typically in drive voltages of several volt and in energy consumptions in the order of pJ/bit. As outlined above, a low energy consumption, however, is highly important to realize a dense integration. Especially in datacenters power consumption of the circuits and their cooling becomes a limiting factor. These all-silicon modulators with their comparatively low efficiency impede signal generation at highest speeds, as electrical driver circuits, which require several volts output swing in order to support high symbol rates, are difficult to realize, and have also a high power consumption.

A different approach to realize phase modulators is the combination of silicon photonics with electro-optic (EO) organic materials, so-called silicon-organic hybrid (SOH) integration. The optical wave is still guided in the silicon waveguide, but interacts with organic material, which has $\chi^{(2)}$-nonlinearity, and offers therefore a pure phase modulation [14–16]. Such Pockels-effect modulators are ideal for Mach-Zehnder and IQ modulators. Using this approach, record low energy consumption could be demonstrated, but so far only for on-off keying and therefore moderate data rates [17,18]. Furthermore, the SOH modulators realized so far are not yet ideally adapted to the standard silicon photonic fabrication. For fabrication they are relying either on e-beam lithography, or on a specially modified DUV lithography process. Also the operation at higher temperatures is a non-trivial issue for organic materials, and investigation of the performance for SOH devices at elevated temperatures has just started.

The work in this thesis addresses those issues of silicon-organic hybrid IQ modulators: By combining custom-tailored organic EO material with optimized silicon slot waveguide structures, highly efficient IQ modulators are realized. Sub-volt drive voltages allow to omit the driver amplifier completely, and the resulting energy consumption for 16QAM modulation at a data rate of 112 Gbit/s reaches a record low value of 19 fJ/bit. This is more than one order of magnitude below state-of-the-art all-silicon modulators, while maintaining a high signal quality with a bit error ratio of 5×10^{-5} [C16,J7]. Using optimized modulator structures, the data rate could be further increased up to 40 GBd 16QAM modulation with an IQ modulator [C21,J9], and 64 GBd 4ASK modulation with an SOH Mach-Zehnder modulator, the currently highest symbol rate for advanced modulation formats generated on the silicon platform [C28,J16]. To start the transition of SOH integration from pure laboratory and research use to a broader scope, the design of the SOH modulator is adapted, and SOH devices are successfully realized on a commercial SOI foundry platform employing 248 nm DUV lithography. Furthermore, organic materials with good efficiency and higher temperature resilience are implemented in SOH devices, and high speed operation of an SOH modulator at 80 °C and ambient atmosphere was demonstrated [C28,J16]. These steps enable the realization of highly complex silicon photonic PICs for next generation data transmission by combining high-performance SOH IQ modulators with the whole portfolio of building blocks, which are available in a commercial silicon photonic foundry.

1.2 Silicon photonic frequency shifter

Not only optical communication benefits from optical integration, but also other applications like distance metrology, vibrometry or heterodyne interferometry can leverage all the advantages of photonic integrated circuits. Especially interferometric devices such as laser Doppler vibrometers can profit from the increased robustness which comes when shrinking the size of such devices [19,20]. Also here, silicon photonics is an ideal platform to realize such PICs. Again, the fabrication in silicon photonic foundries and the availability of a large variety of building blocks create the opportunity to realize PICs for a variety of applications without the need to maintain an expensive fabrication line. However, frequency shifters with a high bandwidth and high side-mode suppression ratio are still missing in the portfolio of silicon photonics. Frequency shifters can be realized on PICs using phase modulators with either serrodyne modulation [21] or single-sideband (SSB) modulation [22]. The amplitude-phase coupling associated with typical all-silicon phase shifters based on free carrier modulation leads, however, to additional side modes which deteriorate the side-mode suppression ratio (SMSR). Current realizations of frequency shifters on silicon rely therefore on thermo-optic phase modulators which offer a pure phase shift, but the bandwidth and therefore the achievable frequency shift of such devices is limited to few tens of kHz [23,24].

To overcome those issues, a frequency shifter based on silicon-organic hybrid integration is realized in this work. The pure phase modulation achieved by the SOH devices is optimal for a high SMSR. Furthermore, because SOH modulators have a large bandwidth, frequency shifts up to 10 GHz could be demonstrated, orders of magnitude larger than former demonstrations of frequency shifters on silicon [C19]. The SSB modulation used in this experiment typically needs a trade-off between conversion efficiency (CE) and SMSR, due to the limited linearity of the Mach-Zehnder transfer function. By using temporal shaping of the electrical drive signal both, a high CE of -5.8 dB and a high SMSR of 23.5 dB could be demonstrated on the SOH-enhanced silicon platform [C14, J17].

2 Theoretical background

In this chapter the fundamental physical relations and their mathematical representations are described, as far as they are necessary for the understanding of this work.

2.1 Second-order nonlinearities and the electro-optic effect

Nonlinear effects occur when the optical properties of a medium are changed under the influence of an external electro-magnetic field. Within this work second-order nonlinear effects within organic materials are used for the modulation of light. In the following chapter, a focused summary of the existing theory is given. The derivation follows the structure in [25]. An in-depth study of nonlinear effects can be found in the book "Nonlinear Optics" by Boyd [26].

2.1.1 Electric polarization

With an electric field $\vec{E}$ imposed on a bulk material, the response of the material is denoted by the polarization $\vec{P}$. In time domain, the spatially local polarization can be described by a Volterra series [26],

$$\begin{aligned}\vec{P}(\vec{r},t) = \varepsilon_0 &\int_{-\infty}^{+\infty} \underline{\chi}^{(1)}(\vec{r},t-t')\vec{E}(\vec{r},t')\mathrm{d}t' \\
+\varepsilon_0 &\int_{-\infty}^{+\infty}\int_{-\infty}^{+\infty} \underline{\chi}^{(2)}(\vec{r},t-t',t-t''):\vec{E}(\vec{r},t')\vec{E}(\vec{r},t'')\mathrm{d}t'\mathrm{d}t'' \\
+\varepsilon_0 &\int_{-\infty}^{+\infty}\int_{-\infty}^{+\infty}\int_{-\infty}^{+\infty} \underline{\chi}^{(3)}(\vec{r},t-t',t-t'',t-t'''):\vec{E}(\vec{r},t')\vec{E}(\vec{r},t'')\vec{E}(\vec{r},t''')\mathrm{d}t'\mathrm{d}t''\mathrm{d}t''' \\
+\ldots&,\end{aligned} \quad (2.1)$$

where ε_0 denotes the vacuum permittivity, and $\underline{\chi}^{(n)}(\vec{r},t',t'',...,t^{(n)})$ are the response functions, tensors of rank $n+1$. Since the response of the material is causal it follows that $\underline{\chi}^{(n)}(\vec{r},t',t'',...,t^{(n)}) = \underline{0}$ for $t',t'',t'''... < 0$.

Eq. (2.1) can be decomposed into a sum of the linear and nonlinear contributions to the electrical polarization

$$\vec{P}(\vec{r},t) = \vec{P}^{(1)}(\vec{r},t) + \vec{P}^{(2)}(\vec{r},t) + \vec{P}^{(3)}(\vec{r},t) + ... = \vec{P}^{(1)}(\vec{r},t) + \vec{P}_{\mathrm{NL}}(\vec{r},t). \qquad (2.2)$$

The linear part of the polarization can be written as

$$\vec{P}^{(1)}(\vec{r},t) = \varepsilon_0 \int_0^{+\infty} \underline{\chi}^{(1)}(\vec{r},t-t')\vec{E}(\vec{r},t')\mathrm{d}t', \qquad (2.3)$$

and after Fourier transformation, see Appendix A.1, it reads

$$\vec{\tilde{P}}^{(1)}(\vec{r},\omega) = \varepsilon_0 \underline{\tilde{\chi}}^{(1)}(\vec{r},\omega)\vec{\tilde{E}}(\vec{r},\omega) \qquad (2.4)$$

for angular frequency ω, introducing the first order electrical susceptibility $\underline{\tilde{\chi}}^{(1)}(\omega)$. Due to the causal response of the medium the real and imaginary parts of the susceptibility are linked via the Kramers-Kronig relation [26]. Hence the susceptibility can only be constant $\underline{\tilde{\chi}}^{(1)}(\omega) \approx \underline{\tilde{\chi}}^{(1)}(\omega_0) = \mathrm{const}$ in a small interval $\omega_1 < \omega_0 < \omega_2$. Using a constant susceptibility in the vicinity of ω_0, Eq. (2.3) simplifies to a linear relation between the electric field and the linear susceptibility,

$$\vec{P}^{(1)}(\vec{r},t) = \varepsilon_0 \underline{\tilde{\chi}}^{(1)}(\vec{r},\omega_0)\vec{E}(\vec{r},t). \qquad (2.5)$$

The electric displacement field is defined as

$$\vec{D}(\vec{r},t) = \varepsilon_0 \vec{E}(\vec{r},t) + \vec{P}(\vec{r},t), \qquad (2.6)$$

using Eq. (2.5) and assuming that the linear contribution of the polarization are much larger than the nonlinear contributions, the electric displacement field can be written as

$$\vec{D}(\vec{r},t) = \varepsilon_0 \left(\underline{1} + \underline{\tilde{\chi}}^{(1)}(\vec{r},\omega_0)\right)\vec{E}(\vec{r},t) = \varepsilon_0 \underline{\varepsilon}_r \vec{E}(\vec{r},t), \qquad (2.7)$$

The quantity $\underline{1}$ is the unity tensor, and $\underline{\varepsilon}_r$ is the relative dielectric permittivity tensor. For the case of an isotropic medium the susceptibility and permittivity can be expressed as a scalar quantity $\bar{\varepsilon}_r = 1 + \underline{\tilde{\chi}}^{(1)} = \bar{n}^2$, with $\bar{n}$ as the linear complex refractive index. The real part n of the complex refractive index is related to the change of the phase velocity of an electro-magnetic wave in the medium while the complex part κ denotes the material absorption.

$$\bar{\varepsilon}_r = \varepsilon_r - j\varepsilon_i = (n - j\kappa)^2 = n^2 - \kappa^2 - j2n\kappa = \bar{n}^2$$
$$n = \sqrt{\varepsilon_r}, \quad \text{if } \kappa \approx 0 \qquad (2.8)$$
$$\kappa = \frac{\varepsilon_i}{2n}$$

2.1.2 Nonlinear polarization and Pockels effect

Following Eq. (2.1) and Eq. (2.2), the nonlinear part of the Polarization $\vec{P}_{NL}$ can be written as

$$\vec{P}_{NL}(\vec{r},t) = \varepsilon_0 \int_{-\infty}^{+\infty}\int_{-\infty}^{+\infty} \underline{\chi}^{(2)}(\vec{r}, t-t', t-t'') : \vec{E}(\vec{r},t')\vec{E}(\vec{r},t'') \, dt' dt''$$
$$+ \varepsilon_0 \int_{-\infty}^{+\infty}\int_{-\infty}^{+\infty}\int_{-\infty}^{+\infty} \underline{\chi}^{(3)}(\vec{r}, t-t', t-t'', t-t''') \vdots \vec{E}(\vec{r},t')\vec{E}(\vec{r},t'')\vec{E}(\vec{r},t''') \, dt' dt'' dt''' \qquad (2.9)$$
$$+ \ldots$$

Since we are only interested in second-order nonlinear effects, higher order contributions to the polarization are omitted. Following the derivation in [26] above equation can be translated in frequency domain and the second-order polarization $\vec{P}_i^{(2)}$ is written as

$$\vec{P}_i^{(2)}(\vec{r}, \omega_n + \omega_m) = \varepsilon_0 \sum_{jk} \sum_{(nm)} \tilde{\chi}_{ijk}^{(2)}(\vec{r}, \omega_n + \omega_m; \omega_1, \omega_2) \vec{E}_j(\vec{r}, \omega_n) \vec{E}_k(\vec{r}, \omega_m). \qquad (2.10)$$

$\underline{\tilde{\chi}}^{(2)}$ is the nonlinear susceptibility tensor, the indices i, j, k denote the Cartesian components and the notation (nm) indicates that the sum $(\omega_n + \omega_m)$ is held fixed during summation while the frequency components ω_n and ω_m are varying.

In the special case of the linear electro-optic effect, or Pockels effect, the refractive index of a material changes under the influence of a low frequency electric field with angular frequency $\omega_m \ll \omega_0, \omega_m \approx 0$. For the case of such a low frequency or static electric field and a high frequency field with angular frequency ω_0, and omitting the spatial dependency for readability, the second-order nonlinear polarization can be written as

$$\tilde{P}_i^{(2)}(\omega_0 + \omega_m \approx \omega_0) = 2\varepsilon_0 \sum_{jk} \tilde{\chi}_{ijk}^{(2)}(\omega_0 + \omega_m; \omega_0, \omega_m) \tilde{E}_j(\omega_0) \tilde{E}_k(\omega_m). \quad (2.11)$$

In the following, the Einstein notation for sums is used. With Eq. (2.11), the electric displacement can be expressed as follows [25]:

$$\tilde{D}_i(\omega_0) = \varepsilon_0 \left(\delta_{ij} + \tilde{\chi}_{ij}^{(1)}(\omega_0) \right) \tilde{E}_j(\omega_0) \\ + 2\varepsilon_0 \tilde{\chi}_{ijk}^{(2)}(\omega_0 + \omega_m; \omega_0, \omega_m) \tilde{E}_j(\omega_0) \tilde{E}_k(\omega_m). \quad (2.12)$$

From this equation, the change of permittivity $\Delta\underline{\varepsilon}_r$, which is dependent on the quasi-static electric field can be derived as

$$\Delta\varepsilon_{r,ij} = 2\tilde{\chi}_{ijk}^{(2)} \tilde{E}_k(\omega_m). \quad (2.13)$$

Using the relation $\varepsilon_r + \Delta\varepsilon_r = (n + \Delta n)^2 \approx n + 2n\Delta n$ the change of refractive index due to the linear electro-optic effect can be described with

$$\Delta n_{ij} = \frac{\tilde{\chi}_{ijk}^{(2)}}{n} \tilde{E}_k(\omega_m). \quad (2.14)$$

The second-order susceptibility $\tilde{\chi}_{ijk}^{(2)}$ can be easily translated into the often used electro-optic coefficient r_{ijk} according to [25] with

$$r_{ijk} = -\frac{2\tilde{\chi}_{ijk}^{(2)}}{n^4}. \quad (2.15)$$

The change of refractive index can therefore be written as

$$\Delta n_{ij} = -\frac{1}{2} n^3 r_{ijk} \tilde{E}_k(\omega_m). \quad (2.16)$$

Considering the permutation symmetry of the fields and for media where the Kleinman symmetry is valid the independent tensor elements are reduced and we can use a contracted notation for the indices i,j,k [26]. For the susceptibility and electro-optic coefficient we introduce the matrices $\tilde{\chi}_{il}^{(2)}$ and r_{il}, following

$$\begin{array}{c} jk: \quad 11 \quad 22 \quad 33 \quad 23,32 \quad 13,31 \quad 12,21 \\ l: \quad 1 \quad 2 \quad 3 \quad 4 \quad 5 \quad 6 \end{array}. \quad (2.17)$$

2.2 Electro-optic modulators for advanced modulation formats

For the modulation of light in integrated optical circuits several concepts exist and are exploited in various devices. However, the most flexible modulator concepts are the Mach-Zehnder modulator (MZM) and on its basis the in-phase–quadrature (IQ) modulator. Based on an arrangement of phase modulators which utilize the linear electro-optic effect, the amplitude and phase of an incoming optical carrier can be arbitrarily adjusted. In the following, a brief summary of the MZM and IQ modulator is given, oriented on the description in [27]. Furthermore a short introduction into the theory of advanced modulation formats is given in this chapter.

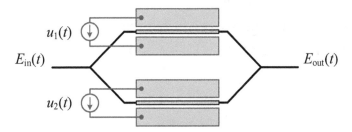

Fig. 2.1: Schematic of a dual drive Mach-Zehnder modulator. The incoming light is split in two arms, each of which is equipped with a phase modulator. After modulation, the light is coherently combined at the output.

2.2.1 Mach-Zehnder modulator

A schematic representation of a MZM is depicted in Fig. 2.1. The incoming optical carrier with the complex electrical field $E_{\text{in}}(t)$ is fed into a Mach-Zehnder interferometer. The carrier is split into two paths, in each arm a phase modulator (PM) is implemented, where an applied voltage $u_{1,2}(t)$ produces a phase shift $\varphi(t)$, proportional to the applied voltage. At the output the arms are combined again and the light can interfere. The electrical field $E_{\text{out}}(t)$ at the output can be described with

$$E_{\text{out}}(t) = \frac{1}{2}\left(e^{j\varphi_1(t)} + e^{j\varphi_2(t)}\right)E_{\text{in}}(t). \tag{2.18}$$

The voltage dependent phase shifts can be described via

$$\varphi_1(t) = \frac{u_1(t)\pi}{U_{\pi,\text{PM1}}} \quad \text{and} \quad \varphi_2(t) = \frac{u_2(t)\pi}{U_{\pi,\text{PM2}}}, \qquad (2.19)$$

where $U_{\pi,\text{PM}}$ is the voltage which has to be applied at one phase modulator to obtain a phase shift of π. Such a configuration of a MZM is called dual-drive operation, because each PM can be operated independently. The MZM can be operated in a push-push mode when the same phase shift is induced in both PMs to modulate only the phase of the output signal. If the phase shift in the two PMs has a relative phase difference of π, i.e. $\varphi_2 = -\varphi_1$, push-pull operation is obtained. The field transfer function of the MZM can then be written as

$$\frac{E_{\text{out}}}{E_{\text{in}}} = \frac{1}{2}\left(e^{j\varphi_1(t)} + e^{-j\varphi_1(t)}\right) = \cos\left(\frac{\Delta\varphi_{\text{MZM}}(t)}{2}\right), \qquad (2.20)$$

where $\Delta\varphi_{\text{MZM}}(t) = \varphi_1(t) - \varphi_2(t) = 2\varphi_1(t)$ is the relative phase difference between both arms of the MZM. Relating this phase shift to a drive voltage we can define

$$\Delta\varphi_{\text{MZM}}(t) = \frac{u(t)\pi}{U_{\pi,\text{MZM}}}. \qquad (2.21)$$

With $u_1(t) = -u_2(t) = u(t)$ we can define $U_{\pi,\text{MZM}}$ as voltage which has to be applied to the MZM in push-pull configuration in order to obtain a relative phase difference of π between the two arms of the MZM, corresponding to $2U_{\pi,\text{MZM}} = U_{\pi,\text{PM}}$ for a configuration as shown in Fig. 2.1. As can be seen from Eq. (2.20), push-pull operation of a MZM has no residual phase modulation and provides therefore a chirp-free amplitude modulation. The intensity transfer function of a MZM in push-pull configuration can be obtained by squaring Eq. (2.20),

$$\frac{I_{\text{out}}}{I_{\text{in}}} = \frac{1}{2} + \frac{1}{2}\cos(\Delta\varphi_{\text{MZM}}(t)) = \frac{1}{2} + \frac{1}{2}\cos\left(\frac{u(t)\pi}{U_{\pi,\text{MZM}}}\right). \qquad (2.22)$$

2.2 Electro-optic modulators for advanced modulation formats

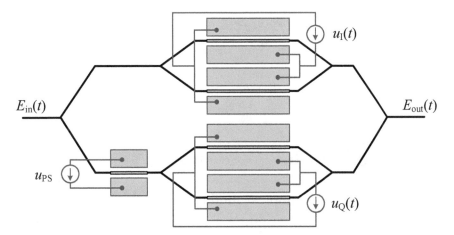

Fig. 2.2: Schematic of an IQ modulator. The incoming carrier is split in two arms. After inducing a relative phase shift of $\pi/2$ using the static voltage u_ps, the in-phase and the quadrature component can be modulated independently by two MZM in push-pull configuration. At the output, both signals are combined. With an IQ modulator each point in the complex plane can be addressed.

Applying a voltage of $u(t) = U_{\pi,\text{MZM}}$ to the MZM in push-pull configuration corresponds to a phase shift of π in the power transfer function, i.e. a transition from constructive interference to destructive interference at the output of the MZM.

2.2.2 In-phase – quadrature modulator

By nesting two MZM an in-phase–quadrature phase (IQ) modulator can be obtained. A schematic of such a modulator is depicted in Fig. 2.2. It consists of two so-called "child" MZM, nested within a "parent" MZM. The incoming light is split and distributed in the in-phase (I) arm of the parent MZM, and in the quadrature (Q) arm. Each arm comprises a MZM that modulates the amplitude of the light. A relative phase shift of $\pi/2$ between I and Q is induced by a static phase shifter. Due to this static phase shift, the I and Q components of the carrier are orthogonal. After modulation both arms are combined at the output. Since the I and Q component of the carrier can be addressed independently in such a structure, every point in the complex plane can be reached. Both child MZM are driven in single-drive push-pull operation and the static phase shifter is set such that a $\pi/2$ phase shift is achieved via $u_\text{PS} = U_{\pi,\text{PM}}/2$. The field transfer function of the IQ modulator can therefore be written as

13

$$\frac{E_{\text{out}}}{E_{\text{in}}} = \frac{1}{2}\cos\left(\frac{\Delta\varphi_{\text{I}}(t)}{2}\right) + j\frac{1}{2}\cos\left(\frac{\Delta\varphi_{\text{Q}}(t)}{2}\right)$$
$$= \frac{1}{2}\cos\left(\frac{u_{\text{I}}(t)\pi}{2U_{\pi,\text{MZM}}}\right) + j\frac{1}{2}\cos\left(\frac{u_{\text{Q}}(t)\pi}{2U_{\pi,\text{MZM}}}\right). \tag{2.23}$$

2.2.3 Advanced modulation formats

On-off keying (OOK) is the simplest modulation format. Binary information is transmitted with every symbol. "Light on" denotes a logical "1", and "light off" a logical "0". However, to achieve a higher spectral efficiency, it is favorable to transmit more than one bit per symbol [10]. This can be achieved by using more than two amplitude levels and/or more than two phase states per symbol. As described in Section 2.2.2, the IQ modulator can address arbitrary phase and amplitude states in the complex plane, making it an ideal tool to generate advanced modulation signals.

Following the notation in [27], for advanced modulation formats each symbol b_k can comprise m bits, $[b_{1,k}, b_{2,k}...b_{m,k}]$. Furthermore, the symbol b_k can be described by a complex phasor with $b_k = b_k^i + jb_k^q$. All symbols form an alphabet A with $M = 2^m$ elements. The symbols are transmitted with the symbol rate $r_S = 1/T_S$ where T_S is the symbol duration.

By utilizing only one MZM, M-ary amplitude shift keying (ASK) can be generated. Being a one-dimensional multi-level signal, it uses only the in-phase component of the complex symbol $b_k = b_k^i$. To obtain the largest spacing between the symbol states in the constellation diagram and thereby the largest signal-to-noise ratio (SNR), the MZM is biased at the null-point, while the m binary bits are defined to be symmetric around 0, and are normalized to unity. With $M/2$ amplitude states and two phase states (0° and 180°), M symbol states can be generated,

$$b_{1,k} \in \left\{-\frac{1}{(M-1)}, \frac{1}{(M-1)}\right\}, b_{2,k} \in \left\{-\frac{3}{(M-1)}, \frac{3}{(M-1)}\right\},$$
$$... b_{m,k} \in \left\{-\frac{(M-1)}{(M-1)}, \frac{(M-1)}{(M-1)}\right\}. \tag{2.24}$$

2.2 Electro-optic modulators for advanced modulation formats

For a single-drive MZM the electrical drive signal is thereby defined as

$$u(t) = U_{\pi,\mathrm{MZM}} + U_D \sum_k b_k \mathrm{rect}\left(\frac{t - kT_S}{T_S}\right), \qquad (2.25)$$

where each symbol b_k is defined for non-return to zero (NRZ) signaling within the symbol duration T_S and U_D is the maximum drive amplitude. Exemplarily the bits of a bipolar 4ASK signal with 4 equidistant levels are defined as

$$b_{1,k} \in \{-1/3, 1/3\} \quad \text{and} \quad b_{2,k} \in \{-1, 1\}. \qquad (2.26)$$

The constellation diagram of such a bipolar 4ASK signal can be seen in Fig. 2.3(a). Since it is symmetric around 0 it can also be described as combination of a binary ASK signal with a binary phase shift keying (PSK) signal, hence it can also be called 2ASK-BPSK. To increase the distance between the symbol states further, one can utilize the full complex plane. Quadrature phase shift keying (QPSK), also called 4-state quadrature amplitude modulation (4QAM) is the most basic modulation format using I and Q component of the carrier. QPSK signals can in principle be generated with one phase modulator or MZM in push-push configuration when using a multi-level drive signal similar to 4ASK signals. However, when utilizing an IQ modulator only two streams of binary signals are needed. The symbols $b_k = b_k^i + jb_k^q$ can be described by

$$b_{1,k}^i = b_{1,k}^q \in \{-1, 1\}. \qquad (2.27)$$

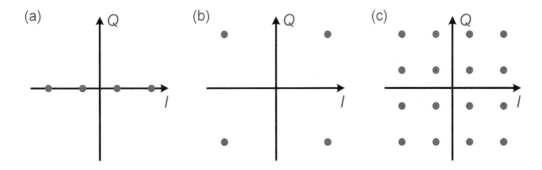

Fig. 2.3: Exemplary constellation diagrams for advanced modulation formats. (a) Constellation diagram of a 4ASK signal, only the in-phase component of the carrier is modulated with a multi-level signal. (b) Constellation diagram of a QPSK signal. The in-phase and quadrature component are each modulated with a binary signal. (c) Constellation diagram of a 16QAM signal, the in-phase and quadrature component of the carrier is modulated with a multilevel signal.

The drive signals for the IQ modulator are then consequently

$$u_\mathrm{I}(t) = U_{\pi,\mathrm{MZM}} + U_D \sum_k b_k^i \mathrm{rect}\left(\frac{t - kT_S}{T_S}\right),$$

$$u_\mathrm{Q}(t) = U_{\pi,\mathrm{MZM}} + U_D \sum_k b_k^q \mathrm{rect}\left(\frac{t - kT_S}{T_S}\right).$$

(2.28)

The constellation diagram of a QPSK signal is depicted in Fig. 2.3(b). Higher-order QAM signals such as 16QAM or 64QAM can be also generated with an IQ modulator by using multi-level drive signals for the I and Q component of the IQ modulator. Corresponding to the generation of bipolar M-ary ASK signals each symbol component b_k^i and b_k^q can represent n bits and since the I and Q component are independent, all combinations are possible, leading to $2^{2n} = 2^m = M$ symbol states which are represented in the complex plane. For the case of 16QAM signals, the bits are defined in symbol states with equidistant amplitudes

$$\begin{aligned} b_{1,k}^i \in \{-1/3, 1/3\} \quad \text{and} \quad b_{2,k}^i \in \{-1, 1\}, \\ b_{1,k}^q \in \{-1/3, 1/3\} \quad \text{and} \quad b_{2,k}^q \in \{-1, 1\}. \end{aligned}$$

(2.29)

For the generation of 16QAM signals, Eq. (2.28) still holds, however, electrical multilevel signals have now to be used instead of binary signals. In Fig. 2.3(c) a constellation diagram of a 16QAM signal is depicted. The constellation points are arranged on a regular grid. Besides using electrical multi-level symbols driving a single IQ modulator, 16QAM signals can also be generated from binary driving signals using, e.g., four nested MZM or an IQ modulator in combination with additional phase modulators. The impact of the different transmitter implementations on the signal quality of the generated 16QAM signals is discussed in detail in [27,28].

3 Energy efficient high-speed IQ modulators

3.1 Silicon-organic hybrid integration

The silicon-on-insulator platform itself does not comprise pure Pockels-effect modulators since silicon as an inversion symmetric crystal does not feature any $\chi^{(2)}$-nonlinearities. The inversion symmetry can be broken by deforming the silicon lattice by overgrowing silicon waveguides with a straining layer of silicon nitride and thus second-order nonlinearities are induced [29,30]. However, the measured effects so far were too small for practical high-speed devices [31]. A way to overcome these limitations of the silicon platform is the combination of standard silicon slot waveguides with organic electro-optic materials [14,32,33]. Silicon forms a passive waveguide, while the geometry of the waveguides is chosen such that the optical wave can interact with the nonlinear electro-optic material. This silicon-organic hybrid (SOH) integration opens the path towards modulators operating at high speed and with a low power consumption. The following section describes the details of the SOH concept for electro-optic modulators, and the realization of Mach-Zehnder and IQ modulators on the basis of SOH integration. Furthermore, the bandwidth of different modulator concepts is discussed as well as different organic electro-optic materials and their stability.

3.1.1 Concept of resistively coupled SOH modulators

The SOH phase modulator as basic building block for different modulator structures is based on a laterally extended slotted waveguide, which is infiltrated with an organic electro-optic cladding material. The cross section of an SOH phase shifter is depicted in Fig. 3.1. Silicon rails form the core of the slot waveguide, the buried oxide (BOX) constitutes the lower cladding, while organic electro-optic material is used for the upper cladding, also infiltrating the slot. Thin, lightly doped conductive silicon slabs connect the rails to metal electrodes. Any voltage applied to these electrodes will therefore drop predominantly in *x*-direction across the slot, leading to a high field intensity of the modulation field inside the slot. A plot of the E_x component of the modulation field can be seen in Fig. 3.1(b). At the same time, the geometry of the slot waveguide leads to an enhancement of the optical electric field inside the slot when operating the waveguide in quasi-TE polarization. This results in a strong interaction of the modulating field with the optical mode in the slot, within the electro-optic material, leading to a highly efficient phase modulator.

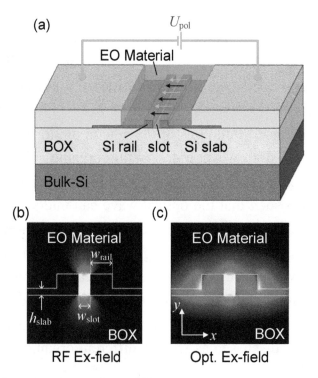

Fig. 3.1: Cross section of an SOH phase shifter. (a) The slot waveguide consists of two rails (rail width w_{rail}, slot width w_{slot}), which are electrically connected to metal electrodes via thin, lightly doped silicon slabs (height h_{slab}). The upper cladding is formed by organic electro-optic material which fills also the slot. To align the chromophores of the electro-optic material, a poling voltage U_{pol} can be applied to the electrodes at elevated temperatures. The molecules (black arrows) align along the electrical field (green arrows). After cooling down and removing the voltage, the chromophores are locked in this alignment. (b) Plot of the dominant E_x component of the RF modulation field. Any applied voltage drops predominantly across the slot leading to a high field intensity in the slot. (c) Plot of the dominant E_x component of the optical field. When operating the waveguide in quasi-TE polarization the discontinuity at the interface between silicon and the organic material leads to an enhancement of the field in the slot.

To achieve a bulk nonlinearity of the electro-optic material, the chromophores have to be aligned in acentric order. After deposition of the organic material, the individual chromophores are randomly oriented, leading to a vanishing bulk nonlinearity. By a so-called poling process, the individual chromophores are reoriented, preferably in acentric order, to achieve a large bulk nonlinearity, i.e. a high r_{33} value. For SOH modulators, the poling process can be done in-situ. After deposition of the material, the device is heated to the glass transition temperature of the organic material. A voltage applied to the metal electrodes of the device leads to an alignment of the now mobile dipoles along the field lines. Within the slot waveguide the dominant field component is oriented in x-direction across the slot, leading to an acentric orientation of the chromophores in x-direction. After cooling down and removing the poling voltage, the molecules are immobilized again and the alignment of the chromophores is locked. After poling a bulk nonlinearity with a significant r_{33} value can be observed in the SOH phase modulator which can be used for Pockels-effect modulation by changing the refractive index with an applied electrical field.

For a propagating wave in a single mode waveguide the change in propagation constant $\delta\beta$ can be described by

$$\delta\beta = \frac{\omega\varepsilon_0}{4\mathcal{P}} \iint_{-\infty}^{\infty} \Delta\varepsilon(x,y)\vec{\mathcal{E}}(x,y)\cdot\vec{\mathcal{E}}^*(x,y)\,\mathrm{d}x\,\mathrm{d}y, \qquad (3.1)$$

where $\mathcal{P}$ is the power in the mode, ω the frequency of the optical signal, $\vec{\mathcal{E}}(x,y)$ is the transversal electric mode field of the optical signal and $\Delta\varepsilon(x,y)$ a change of the permittivity [25,34]. Using Eq. (2.16) the permittivity change in the case of a Pockels-effect modulator can be written as

$$\Delta\varepsilon(x,y) = 2n_{\mathrm{EO}}\Delta n(x,y) = -n_{\mathrm{EO}}^4 r_{33}(x,y) E(x,y,\omega_m), \qquad (3.2)$$

where r_{33} is the bulk electro-optic coefficient, n_{EO} the refractive index of the nonlinear material and $E(x,y,\omega_m)$ is the electric modulation field at the microwave frequency ω_m. The phase change $\delta\varphi = -\delta\beta L$ within an SOH modulator of length L can therefore be written as

$$\begin{aligned}\delta\varphi &= \frac{\omega\varepsilon_0 L}{4\mathcal{P}} \iint_{-\infty}^{\infty} n_{\mathrm{EO}}^4 r_{33}(x,y) E(x,y,\omega_m) \vec{\mathcal{E}}(x,y)\cdot\vec{\mathcal{E}}^*(x,y)\,\mathrm{d}x\,\mathrm{d}y \\ &= \frac{\omega\varepsilon_0 L}{4\mathcal{P}} \iint_{-\infty}^{\infty} n_{\mathrm{EO}}^4 r_{33}(x,y) E(x,y,\omega_m) |\vec{\mathcal{E}}(x,y)|^2 \,\mathrm{d}x\,\mathrm{d}y.\end{aligned} \qquad (3.3)$$

As it can be seen in Eq. (3.3), the magnitude of the phase shift depends on material parameters such as the refractive index n_{EO} and the electro optic coefficient r_{33}, as well as on the strength of the modulation field $E(x,y,\omega_m)$ and the overlap of the optical mode field and the modulation field.

If we approximate the slot as a parallel plate capacitor, we can use a homogenous field with constant frequency ω_m within the area of the slot A_{slot} and a homogenous r_{33}

$$E(x,y,\omega_m) = \begin{cases} E_m & \forall x,y \in A_{slot} \\ 0 & \text{else} \end{cases}. \quad (3.4)$$

$$r_{33}(x,y) = r_{33}$$

Using this, Eq. (3.3) can be simplified within the area of the slot. By a defining a field interaction factor Γ with

$$\Gamma = \frac{n_{EO,slot}\sqrt{\varepsilon_0}}{\sqrt{\mu_0}} \frac{\iint_{A_{slot}} |\vec{\mathcal{E}}(x,y)|^2 \, dx\, dy}{\iint \text{Re}\{\vec{\mathcal{E}}(x,y) \times \vec{\mathcal{H}}(x,y)\} \cdot e_z \, dx\, dy}, \quad (3.5)$$

Eq. (3.3) now reads

$$\delta\varphi = \frac{\omega \varepsilon_0 n_{EO,slot}^3 r_{33} \sqrt{\mu_0} \Gamma E_m L}{2\sqrt{\varepsilon_0}} = \frac{\pi n_{EO,slot}^3}{\lambda} r_{33} \Gamma E_m L, \quad (3.6)$$

where λ is the free-space wavelength of the optical signal. With the parallel plate assumption, we can further approximate that the modulating field is generated by an applied voltage with $E_m = U_m / w_{slot}$. We can then define the π-voltage $U_{\pi,PM}$ as the voltage which has to be applied to the phase modulator in order to achieve a phase shift of $\delta\varphi = \pi$

$$U_{\pi,PM} = \frac{w_{slot} \lambda}{\Gamma n_{EO,slot}^3 r_{33} L}. \quad (3.7)$$

Eq. (3.7) gives a comprehensive insight on the influence of different material and geometry parameters on the modulator. The field interaction factor Γ depends strongly on the geometry of the slot waveguide. By evaluating Eq. (3.5) with field values obtained from numerical simulations for a parameter set of w_{slot}, w_{rail} and h_{slab} optimal geometries for a maximized Γ can be obtained for SOI slot waveguides with a given height of the silicon device layer

$h_{Si} = 220$ nm [25]. In Fig. 3.2 Γ is plotted for different slab heights and slot widths. The rail width is held at $w_{rail} = 240$ nm which is approximately the optimum for a device height of $h_{Si} = 220$ nm. It can be seen from the graph that both, a smaller slot width and a thinner slab height lead to a higher field confinement. However, a slot width of approx. $w_{slot} = 80$ nm is the fabrication limit with E-beam technology and approx. $w_{slot} = 120$ nm is the smallest slot width which can be reliably fabricated with optical lithography. Furthermore, there is a trade-off between bandwidth and efficiency. As it will be discussed in the next section, the capacity of the slot, together with the resistivity of the slab dictates the bandwidth of the SOH modulator. By decreasing the slot width the capacity is increased and by decreasing the slab height the resistivity is increased. Both lead to a lower bandwidth of the modulator.

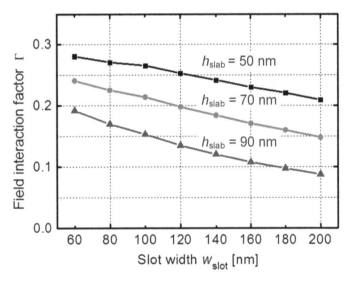

Fig. 3.2: Field interaction of the optical mode with the EO material in the slot A_{slot} as function of slot width at a wavelength of 1550 nm. The rail width was fixed at $w_{rail} = 240$ nm which is approx. the optimum value for a height of the silicon layer of $h_{Si} = 220$ nm. Thin slabs and narrow slots lead to a high field interaction factor, however, it has to be balanced with bandwidth requirements of the modulation.

While using parallel plate approximation is very intuitive and easy to implement in simulation, it has only limited accuracy. For a more exact description we can modify the field interaction factor Γ. By introducing a weighting factor

$$\gamma(x,y) = \begin{cases} \dfrac{E(x,y)}{E_m} & \forall x,y \in A_{\mathrm{EO}} \\ 0 & \forall x,y \notin A_{\mathrm{EO}} \end{cases} \qquad (3.8)$$

and using Eq. (3.5) we can write

$$\Gamma_\gamma = \frac{n_{\mathrm{EO}}}{Z_0} \frac{\iint \left|\vec{\mathcal{E}}(x,y)\right|^2 \gamma(x,y)\,dx\,dy}{\iint \mathrm{Re}\left\{\vec{\mathcal{E}}(x,y) \times \vec{\mathcal{H}}(x,y)\right\} \cdot e_z\,dx\,dy}. \qquad (3.9)$$

Thereby we weight each point within the electro-optic material with its RF field, normalized to the homogenous field of the parallel plate approximation. Since the information on the spatial dependent field values is represented within the weighted field interaction factor Γ_γ, the compact Eq. (3.7) for the π-voltage can still be used.

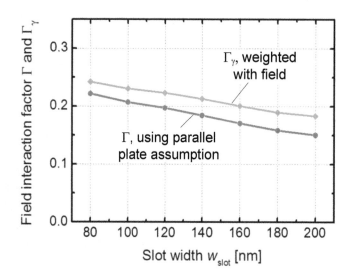

Fig. 3.3: Comparison between field interaction factor Γ and weighted field interaction factor Γ_γ for different slot widths. The slab height is $h_{\mathrm{slab}} = 70\,\mathrm{nm}$, the rail width $w_{\mathrm{rail}} = 240\,\mathrm{nm}$ and the device height $h_{\mathrm{Si}} = 220\,\mathrm{nm}$. The weighted field interaction factor shows approximately 10% higher value than the field interaction factor calculated only with parallel plate assumption.

3.1 Silicon-organic hybrid integration

In Fig. 3.3 the field interaction factors Γ_γ and Γ are plotted for different slot widths. It can be seen that the weighted interaction factor Γ_γ is approximately 10% larger than the interaction factor which is based on the parallel plate approximation. This validates the parallel plate approximation, however, for an precise determination of the in-device r_{33}, which is derived from the measured U_π, it is beneficial to use the weighted field interaction factor Γ_γ.

With the SOH phase modulators as basic building block, Mach-Zehnder modulators and IQ modulators can easily be realized on the SOI platform. In Fig. 3.4 a schematic of an IQ modulator is depicted. It consists of two nested Mach-Zehnder modulators, each of which contains two SOH phase modulators, as discussed in Section 2.2. Standard SOI strip waveguides are used to transport the optical signal. To split and combine the light, 2x2 multimode interference coupler (MMI) are used [35]. The transfer matrix for the optical field at the 2x2 MMI can be written as

$$\mathbf{M}_{2\times 2} = \frac{1}{\sqrt{2}} \begin{pmatrix} 1 & -j \\ -j & 1 \end{pmatrix}. \tag{3.10}$$

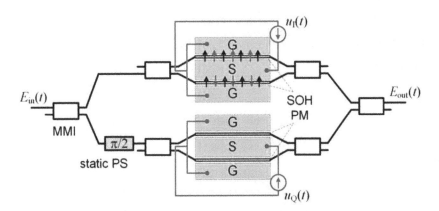

Fig. 3.4: Schematic of an SOH IQ modulator. Two MZM each with two SOH PM form the IQ modulator. Standard SOI strip waveguides are used as access waveguides and MMIs are used to split and combine the optical signal. The $\pi/2$ phase difference between the I and Q arm is introduced by a static phase shift (PS). The SOH PM are connected to a transmission line in ground-signal-ground (GSG) configuration. Push-pull operation is achieved by applying a poling voltage across the floating ground electrodes of the MZM. Thereby the chromophores (black arrows) align in both arms in the same direction, and when applying a modulation signal to the GSG line, the electrical field (blue arrows) is in one arm parallel, in the other arm antiparallel oriented.

To obtain the necessary $\pi/2$ phase shift between the I and Q arm of the modulator, an additional static phase shift is introduced in one arm of the IQ structure, as depicted in Fig. 3.4. The SOH MZM are implemented as single drive MZM in push pull configuration, each MZM is connected to a transmission line in ground-signal-ground (GSG) configuration. The push-pull operation is realized by poling the MZM via the floating ground electrodes. Thereby the chromophores are aligned in both arms in the same direction (black arrows in Fig. 3.4). When applying a modulation signal to the GSG the electric field (blue arrows) is in one arm parallel, in the other arm antiparallel to the alignment of the chromophores, generating a push-pull signal.

3.1.2 Capacitively coupled SOH modulators

The SOH integration relies on the strong overlap of the optical field and the modulation field in the organic electro-optic material. For the resistive coupled SOH modulator described in Section 3.1.1 the optical field is confined to the slot due to the refractive index contrast of the silicon waveguide, while the confinement of the electrical wave is due to the resistive coupling of the rails via doped silicon slabs to the metal electrodes. While this configuration ensures a high field intensity in the slot which is filled with EO material, the resistive slab has the disadvantage that conductive areas have to be in close proximity to the guided optical wave. This can lead to an increase of the optical losses, furthermore, the bandwidth of the device is limited by the slab resistance, as explained in the next section. As alternative to the resistive slab, the modulation field can be confined to the slot via dielectric materials with high permittivity at the modulation frequency ω_m, so called high-K materials. In Fig. 3.5 a cross section of a capacitive coupled slot waveguide is depicted. A silicon slot waveguide on the SOI platform defines the core of the optical waveguide, the buried oxide defines the lower cladding while organic electro-optic material defines the upper cladding of the waveguide and infiltrates the slot. Metal electrodes are defined laterally along the waveguide, outside of the influence of the optical field. Between metal electrodes and the silicon rails high-K material is deposited. The material is chosen such, that the refractive index for the optical wave is smaller than the index of the core. E.g. titanium dioxide or barium titanate, have refractive indices of $n = 2.1...2.5$ at a wavelength of $\lambda = 1550\,\text{nm}$, while the permittivity is $\varepsilon_r = 100...500$ for RF frequencies in the GHz-range. The discontinuity at the interface between the silicon and the EO material leads to a field enhancement of the optical mode within the slot. Similarly, the discontinuity at the interface between the high-K material and the silicon slot waveguide leads to an enhancement of the modulation field within the slot. In a lumped element approximation, the device can be modeled as a slot capacitance C_{slot}, which is coupled to the electrodes via the coupling capacitance $C_{\text{high-K}}$, defined by the material with high relative permittivity, see Fig. 3.5(a).

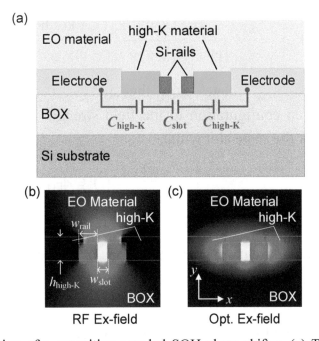

Fig. 3.5: Cross section of a capacitive coupled SOH phase shifter. (a) The slot waveguide consists of two rails (rail width w_{rail}, slot width w_{slot}). Between the electrodes and the rails is material with high permittivity for RF frequencies (high-K) deposited, which has a lower refractive index at the wavelength of the optical carrier. The upper cladding is formed by organic electro-optic material which fills also the slot. (b) Plot of the dominant E_x component of the RF modulation field. Due to the discontinuity of the permittivity, any applied voltage drops predominantly across the slot leading to a high field intensity in the slot. The silicon rails are undoped and modeled as dielectric. For the high-K material a permittivity of $\varepsilon_r = 200$ is used. (c) Plot of the dominant E_x component of the optical field. When operating the waveguide in quasi-TE polarization the discontinuity at the interface between silicon and the organic material leads to an enhancement of the field in the slot, the refractive index of the organic material is modeled with $n_{\text{EO}} = 1.7$, the material with high permittivity with $n_{\text{high-K}} = 2.3$.

If the material exhibits a large enough permittivity, we can neglect the contribution of $C_{\text{high-K}}$, placed in series to the slot capacitance C_{slot}. In that case any applied RF signal drops mainly across the slot of the slot waveguide, leading to a high overlap between the optical field and the modulation field. A plot of the dominant E_x components of the RF field and optical field is depicted in Fig. 3.5(b) and (c).

As described in Section 3.1.1, an important figure of merit is the field interaction factor Γ. Also for the capacitively coupled SOH modulator we can define a weighted field confinement factor $\Gamma_{\gamma_{cc}}$ as

$$\Gamma_{\gamma_{cc}} = \frac{n_{\text{EO}}}{Z_0} \frac{\iint |\vec{\mathcal{E}}(x,y)|^2 \gamma_{cc}(x,y) \, dx \, dy}{\iint \text{Re}\{\vec{\mathcal{E}}(x,y) \times \vec{\mathcal{H}}(x,y)\} \cdot e_z \, dx \, dy}, \qquad (3.11)$$

the weighting factor γ_{cc} relates now the strength of the modulation field $E_x(x,y)$ for a given configuration to the field strength in a parallel plate approximation E_m for the same slot width,

$$\gamma_{cc}(x,y) = \begin{cases} \dfrac{E(x,y)}{E_m} & \forall x,y \in A_{\text{EO}} \\ 0 & \forall x,y \notin A_{\text{EO}} \end{cases}. \qquad (3.12)$$

This ensures that the field interaction factor of the resistively coupled slot waveguide and the field interaction factor for a capacitively coupled slot waveguide can be directly compared.

Besides the geometrical considerations like slot width, which are discussed in Section 3.1.1, the dielectric properties of the high-K material are of importance for the capacitive coupled SOH modulator: The refractive index at the optical wavelength of $\lambda = 1550$ nm influences the confinement of the optical mode in the slot, while the relative permittivity at RF frequencies in the GHz regime defines the confinement of the modulation field. In Fig. 3.6 the calculated field interaction factor is plotted for different refractive indices at $\lambda = 1550$ nm and different relative permittivities at $f_m = 20$ GHz. A low refractive index leads to a high confinement of the optical mode within the slot, thereby leading to a high field interaction factor. At the same time a high relative permittivity at the modulation frequency leads to a strong confinement of the modulation field within the slot and thereby to also to a large field confinement. For an efficient modulator, a material is needed which combines those two properties. An example of such a material would be $Ba_xSr_{1-x}TiO3$ with a refractive index of $n_{1550nm} = 2.0$ [36] and having a permittivity of more than $\varepsilon_r = 300$ [37].

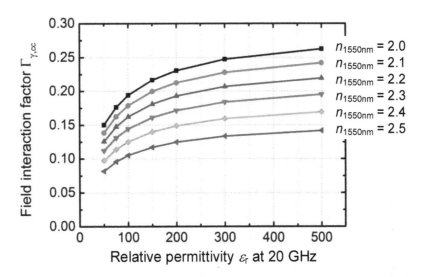

Fig. 3.6: Calculated field interaction factor $\Gamma_{\gamma_{cc}}$ of a capacitively coupled SOH phase modulator with for different dielectric parameter of the high-K material. The rail width is $w_{rail} = 240\,\text{nm}$, the slot width is $w_{slot} = 120\,\text{nm}$ and the height of the high-K material is $h_{high-K} = 300\,\text{nm}$. The amplitude of the modulation field is normalized to the parallel-plate approximation, therefore the values are dircetly comparable to the values of Γ_γ in Fig. 3.3, for 120 nm slotwidth.

3.1.3 Bandwidth of SOH modulators

The bandwidth of the SOH modulator is predominantly influenced by the RC limitation of the slot capacitance, which is charged and discharged via the electrodes, by the losses of the transmission line, and by the walk off between the optical and electrical wave in a travelling wave configuration. In the following section the bandwidth of the device is derived according to the lumped element model in [25] and extended to capacitively coupled modulators.

3.1.3.1 Bandwidth of a resistive coupled SOH modulator

As depicted in Fig. 3.1, the slot of a resistively coupled SOH phase modulator forms a capacitance C_{slot} which is charged and de-charged via the thin silicon slabs with resistance R_{slab}. For small modulation frequencies, the SOH device can be treated as lumped element. For terminated operation, the equivalent circuit diagram is depicted in Fig. 3.7.

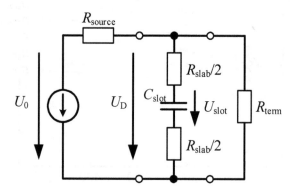

Fig. 3.7: Equivalent circuit of an SOH phase modulator in a lumped element approximation. The source is described with the open loop voltage U_0 and the source resistance R_{source}. The modulator is modeled as RC element with the resistance R_{slab} for both silicon slabs and the capacitance of the slot C_{slot}. The device is terminated with R_{term}.

Since the modulation dependents on the field within the slot, the phase shift is proportional to the voltage across the slot waveguide $\delta\varphi \propto U_{slot}(\omega_m)$. According to the equivalent circuit in Fig. 3.7 it can be expressed as

$$|U_{slot}| = |U_D(\omega_m)| \left| \frac{Z_{slot}(\omega_m)}{R_{slab} + Z_{slot}(\omega_m)} \right|$$

$$= U_0 \frac{\left|(R_{slab} + Z_{slot}(\omega_m))R_{term}\right|}{\left|(R_{slab} + Z_{slot}(\omega_m))R_{term} + (R_{slab} + Z_{slot}(\omega_m) + R_{term})R_{source}\right|} \quad (3.13)$$

$$\cdot \left| \frac{Z_{slot}(\omega_m)}{R_{slab} + Z_{slot}(\omega_m)} \right|.$$

3.1 Silicon-organic hybrid integration

For a matched system with $R_{\text{load}} = R_{\text{source}}$ Eq. (3.13) can be written as

$$|U_{\text{slot}}| = U_0 \frac{|R_{\text{slab}} + Z_{\text{slot}}(\omega_m)|}{|2R_{\text{slab}} + 2Z_{\text{slot}}(\omega_m) + R_{\text{term}}|} \cdot \frac{|Z_{\text{slot}}(\omega_m)|}{|R_{\text{slab}} + Z_{\text{slot}}(\omega_m)|}$$
$$= U_0 \frac{|Z_{\text{slot}}(\omega_m)|}{|2R_{\text{slab}} + 2Z_{\text{slot}}(\omega_m) + R_{\text{term}}|}. \quad (3.14)$$

The frequency dependent phase shift for one phase modulator can therefore be written as

$$\delta\varphi(\omega_m) = \delta\varphi_0 \frac{2|Z_{\text{slot}}(\omega_m)|}{|2R_{\text{slab}} + 2Z_{\text{slot}}(\omega_m) + R_{\text{term}}|}$$
$$= \delta\varphi_0 \frac{1}{\left|1 + \dfrac{R_{\text{slab}} + R_{\text{term}}/2}{Z_{\text{slot}}(\omega_m)}\right|}. \quad (3.15)$$

A factor of 2 is introduced so that for a modulation frequency of $\omega_m = 0$ the phase shift is $\delta\varphi(0) = \delta\varphi_0$. As it can be seen in Eq. (3.15) the bandwidth of the SOH phase shifter is limited by the finite resistivity of the silicon slabs and the source and termination resistance of the RF system.

Additional limitations are the velocity mismatch of the electrical and optical wave, as well as the microwave loss on the transmission line. For the frequency dependent losses of the transmission line the loss parameter $\alpha(\omega_m)$ can be used, comprising the Ohmic and dielectric power loss in the transmission line. An effective length is introduced, it is defined as the length L_{eff} of an phase matched and lossless transmission line for a phase modulator, which gives the same modulation as a lossy transmission line of length L [38].

$$L_{\text{eff}}(\omega_m) = \int_0^L e^{-\frac{\alpha(\omega)}{2}z} \, dz = \frac{2}{\alpha(\omega)}\left(1 - e^{-\frac{\alpha(\omega)}{2}L}\right). \quad (3.16)$$

The factor 1/2 in the exponent stems from the fact that $\alpha(\omega_m)$ is a power attenuation factor and for a Pockels-effect modulation the attenuation of the field is of relevance. The velocity mismatch is represented by

$$\varphi(\omega_\mathrm{m},t) = \delta\varphi_0 \frac{1}{L}\int_0^L \cos\left(\omega_\mathrm{m}\left(\frac{\Delta t_{\mathrm{o,m}}}{L}z - t\right)\right)\mathrm{d}z$$
$$= \delta\varphi_0 \frac{\sin(\omega_m t) - \sin(\omega_m(t - \Delta t_{\mathrm{o,m}}))}{\Delta t_{\mathrm{o,m}}\omega_m}, \quad (3.17)$$

where $\Delta t_{\mathrm{o,m}}$ is the temporal walk off between the optical and the electrical wave, described by

$$\Delta t_{\mathrm{o,m}} = \left|\frac{L}{v_{\mathrm{g,o}}} - \frac{L}{v_{\mathrm{g,m}}}\right| = \frac{L}{c}\left|n_{\mathrm{g,o}} - n_{\mathrm{g,m}}\right|, \quad (3.18)$$

where $v_{\mathrm{g,o}}$ and $v_{\mathrm{g,m}}$ is the group velocity of the modulation signal and the optical signal, respectively, and $n_{\mathrm{g,o}}$ and $n_{\mathrm{g,m}}$ the corresponding group index. Taking into account the individual contributions, the frequency dependent phase shift of an SOH phase modulator can be expressed as

$$\varphi(\omega_\mathrm{m},t) = \delta\varphi_0 \frac{1}{\left|1 + \dfrac{R_{\mathrm{slab}} + R_{\mathrm{term}}/2}{Z_{\mathrm{slot}}(\omega_m)}\right|} \frac{L_{\mathrm{eff}}(\omega_m)}{L} \frac{\sin(\omega_m t) - \sin(\omega_m(t - \Delta t_{\mathrm{o,m}}))}{\Delta t_{\mathrm{o,m}}\omega_m}. \quad (3.19)$$

If the temporal walk off can be neglected for small device lengths, Eq. (3.19) can be simplified to

$$\varphi(\omega_\mathrm{m}) = \delta\varphi_0 \frac{1}{\left|1 + \dfrac{R_{\mathrm{slab}} + R_{\mathrm{term}}/2}{Z_{\mathrm{slot}}(\omega_m)}\right|} \frac{L_{\mathrm{eff}}(\omega_m)}{L} \quad (3.20)$$

In the case of a single-drive MZM in push-pull configuration, where two SOH phase shifter are driven in parallel with the same source and termination resistance, the equation changes to

$$\varphi_{\mathrm{MZM}}(\omega_\mathrm{m}) = \delta\varphi_0 \frac{1}{\left|1 + \dfrac{R_{\mathrm{slab}} + R_{\mathrm{term}}}{Z_{\mathrm{slot}}(\omega)}\right|} \frac{L_{\mathrm{eff}}(\omega_m)}{L}. \quad (3.21)$$

The electro-optic response of a MZM, defined by its linear small-signal intensity modulation, can therefore be written as

$$S_{21,\text{EO}}(\omega_m) = 10\log_{10}\left(\frac{\delta I(\omega_m)}{\delta I(0)}\right) = 10\log_{10}\left(\frac{\delta\varphi_{\text{MZM}}(\omega_m)}{\delta\varphi(0)}\right). \quad (3.22)$$

3.1.3.2 Bandwidth of a capacitive coupled SOH modulator

For the case of a capacitively coupled SOH modulator, the resistive slabs are omitted. Instead, a high-K dielectric is introduced between the slot and the electrodes. Similar as in Section 3.1.3.2 the bandwidth of such a structure can be described with a lumped-element approach. The equivalent circuit of a capacitive coupled phase modulator is depicted in Fig. 3.8. Instead of the slab resistance R_{slab}, the capacitance $C_{\text{high-K}}$ couples the drive voltage to the slot.

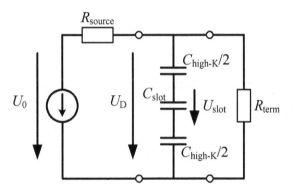

Fig. 3.8: Equivalent circuit of a capacitive coupled SOH phase modulator in a lumped element approximation. The source is described with the open loop voltage U_0 and the source resistance R_{source}. The modulator is modeled as capacitor comprising the capacitance $C_{\text{high-K}}$ for the dielectric and the capacitance of the slot C_{slot}. The device is terminated with R_{term}.

Using the impedance

$$Z_{\text{high-K}} = \frac{1}{j\omega_m C_{\text{high-K}}}, \quad (3.23)$$

the frequency dependent voltage $U_{\text{slot}}(\omega_m)$ can be described as

$$|U_{\text{slot}}(\omega_m)| = U_0 \frac{|Z_{\text{high-K}}(\omega_m) + Z_{\text{slot}}(\omega_m)|}{|2Z_{\text{high-K}}(\omega_m) + 2Z_{\text{slot}}(\omega_m) + R_{\text{term}}|}$$

$$\cdot \frac{|Z_{\text{slot}}(\omega_m)|}{|Z_{\text{high-K}}(\omega_m) + Z_{\text{slot}}(\omega_m)|} \quad (3.24)$$

$$= U_0 \frac{|Z_{\text{slot}}(\omega_m)|}{|2Z_{\text{high-K}}(\omega_m) + 2Z_{\text{slot}}(\omega_m) + R_{\text{term}}|}.$$

Analogous to Eq. (3.15) for $R_{\text{source}} = R_{\text{term}}$, the frequency dependent phase shift for a phase matched phase modulator can be written as

$$\delta\varphi(\omega_m) = \delta\varphi_0 \frac{2|Z_{\text{slot}}(\omega_m)|}{|2Z_{\text{high-K}}(\omega_m) + 2Z_{\text{slot}}(\omega_m) + R_{\text{term}}|} \frac{L_{\text{eff}}(\omega_m)}{L}$$

$$= \delta\varphi_0 \frac{1}{\left|1 + \frac{Z_{\text{high-K}}(\omega_m)}{Z_{\text{slot}}(\omega_m)} + \frac{R_{\text{term}}/2}{Z_{\text{slot}}(\omega_m)}\right|} \frac{L_{\text{eff}}(\omega_m)}{L} \quad (3.25)$$

To achieve a good field confinement for a capacitively coupled SOH modulator, it is beneficial to have a material between the electrodes and the slot with a permittivity at least an order of magnitude higher than for the material within the slot. This leads to a large capacitance $C_{\text{high-K}} \gg C_{\text{slot}}$. Using this assumption Eq. (3.25) reads

$$\delta\varphi(\omega_m) \stackrel{Z_{\text{slot}} \gg Z_{\text{high-K}}}{=} \delta\varphi_0 \frac{1}{\left|1 + \frac{R_{\text{term}}/2}{Z_{\text{slot}}(\omega_m)}\right|} \frac{L_{\text{eff}}(\omega_m)}{L}. \quad (3.26)$$

Analogous for a single drive MZM in push-pull configuration we find

$$\delta\varphi_{\text{MZM}}(\omega_m) \stackrel{Z_{\text{slot}} \gg Z_{\text{high-K}}}{=} \delta\varphi_0 \frac{1}{\left|1 + \frac{R_{\text{term}}}{Z_{\text{slot}}(\omega_m)}\right|} \frac{L_{\text{eff}}(\omega_m)}{L}. \quad (3.27)$$

The bandwidth of a capacitive coupled SOH modulator is only determined by the capacitance of the slot and the source and termination impedance in the feeding network. For a proper choice of the device parameters this leads to a larger EO bandwidth. Furthermore, as the capacitive coupled design can omit the resistive silicon slabs, the Ohmic losses of the transmission line are reduced, which increases the bandwidth of the device further [38].

3.1.4 Organic electro-optic materials

As described in Section 2.1.2, for a Pockels-effect modulator, light must interact with a material with non-vanishing $\underline{\chi}^{(2)}$-tensor. It can be shown that this can only be obtained in materials which have a non-centrosymmetric structure [26], which can be fulfilled in various crystalline structures. Lithium niobate (LiNbO$_3$) is the most prominent example and can provide moderate nonlinearities around $r_{33} \approx 30 \text{ pm/V}$. Besides such crystalline materials, nonlinear organic materials are of highest interest for technical applications. They can provide high nonlinearities up to $r_{33} \approx 500 \text{ pm/V}$, an intrinsic bandwidth of tens of THz [39], and they offer good processability since they can be easily combined with other material systems. In the following section, a brief overview is given on the background and the theory of organic nonlinear materials. The presentation follows the explanation in [39] and [40], which give detailed insight in those materials.

Nonlinear organic molecules, or chromophores, usually consist of a donor and an acceptor part, which is connected via a π-conjugated electron bridge. Such a structure has a high dipole moment and allows an easy polarization of the delocalized electron density. The polarization of a single molecule via an external electric field can be described with the molecular dipole moment $\vec{p}$ which is induced by the electrical field $\vec{E}$. Similar as for the bulk nonlinearity it can be represented by a power series expansion with

$$\vec{p} = \alpha_{ij} E_j + \beta_{ijk} : E_j E_k + \gamma_{ijkl} \vdots E_j E_k E_l, \qquad (3.28)$$

where α_{ij} describes the linear polarizability, β_{ijk} is the first-order and γ_{ijkl} is the second-order hyperpolarizability. The indices $i, j, k = \{x, y, z\}$ give the orthogonal coordinate axis with respect to the molecule. For a donor-acceptor system with a rod-like shape only a single element of the hyperpolarizability tensor is significant. If we define the local coordinate system to such, that the chromophore is elongated along the z-axis, the axis of the dipole moment, the significant tensor element is β_{zzz}. To obtain a second-order nonlinearity in bulk, it is not only necessary to break the symmetry by dipolar, hence asymmetric molecules with a hyperpolarizability β_{zzz}, but also the ensemble of those chromophores has to be non-centrosymmetric. This requires an alignment of the chromophores in so-called acentric order and the second-order susceptibility tensor $\chi^{(2)}_{333}(\omega)$ for bulk can then be written as

$$\chi^{(2)}_{333}(\omega) = N\beta_{zzz}(\omega)\langle\cos^3\theta\rangle g(\omega). \tag{3.29}$$

Here N is the chromophore density, $\langle\cos^3\theta\rangle$ the average acentric order parameter, which describes the orientation of the molecular z-axis with respect to the polarization of the external field. $g(\omega)$ is the Lorentz-Onsager field factor, correcting for partial screening of the applied field. Using Eq. (2.15), the nonlinearity parameter $r_{33}(\omega)$ can be described with

$$r_{33} = -\frac{2\tilde{\chi}^{(2)}_{333}}{n^4} = -\frac{2N\beta_{zzz}(\omega)\langle\cos^3\theta\rangle g(\omega)}{n^4}. \tag{3.30}$$

While the chromophores account for the electro-optic activity, those dipole molecules are normally either embedded into an often polymeric matrix material or are engineered such that molecular side groups prevent direct dipole-dipole interaction [39]. The properties of such a material system are therefore not only determined by the electro-optical core but also to a large extend by the matrix in which it is embedded.

3.1.4.1 Stability of organic electro-optic materials

The temporal and thermal stability of electro-optic material systems is of large interest for practical applications. The bulk nonlinearity parameter r_{33} depends as described in Eq. (3.30), on the chromophore density N and the average acentric order $\langle\cos^3\theta\rangle$, both of which can be strongly dependent on time.

The photon induced degradation of organic EO molecules, or photobleaching, is a main contributor to the reduction of the density of the active chromophores N. The underlying mechanism of photo bleaching is dominated by oxidation of the EO molecules via singlet oxygen which is formed in excited triplet states of the molecules. Following the explanation in [41], the bleaching process can be described as follows: Photons excite the organic molecules from their singlet ground state S_0 into the first excited state S_1. From the excited state S_1 significant inter system crossing into the energetically close triplet state T_1 can occur, from which it decays again with an inter system crossing into the ground state S_0. Those inter system crossings processes $S_1 \rightarrow T_1$ and $T_1 \rightarrow S_0$ interact with the dissolved oxygen, resulting in the formation of singlet oxygen, which is much more reactive than oxygen molecules in their triplet ground states. Organic molecules in the ground state S_0 can now react with the singlet oxygen forming a peroxide and thereby losing the EO properties. Both, high light intensity and the availability of oxygen is needed for the photo bleaching to occur [41,42]. While for experimental investigations often visible light within the absorption band of the EO molecule is used, also near infrared light can contribute to photo bleaching. While infrared light may be outside of the absorption band of the molecule, still excitation into higher states can occur due to two photon absorption, inducing the bleaching process. To obtain a significant contribution of the two photon absorption to the photo bleaching, a high intensity within the organic material is needed. However, in silicon photonic waveguides with very compact cross section high intensities in the range of MW/cm^2 can easily be reached, even when using moderate optical powers in the order of few milliwatt. Therefore, photo bleaching cannot be neglected when considering SOH devices operated at a wavelength around 1550 nm.

In addition to possible photo bleaching, the alignment stability is key for a stable $\langle \cos^3 \theta \rangle$ and hence a stable nonlinearity over time. For chromophores embedded in amorphous matrix materials the thermal properties of the matrix material define the alignment stability. The glass transition temperature T_g is the temperature at which the matrix material undergoes the transition from liquid state into a glassy state. While for poling the material is normally heated at or above the glass transition temperature to facilitate the alignment of the organic molecules, the matrix should be as rigid as possible for operation. A detailed study and modelling of various matrix materials was conducted in [43] with respect to relaxation at various temperatures. The relaxation processes $\psi(t)$ follow typically a stretched exponential function, or Kohlrausch-Williams-Watts function [44] which has the form of

$$\psi(t) = \exp\left[-(t/\tau)^b\right]. \tag{3.31}$$

where τ is the relaxation time constant and $0 < b \leq 1$ the stretching parameter. Using this approach, ideal parameters for a matrix material can be extrapolated: To obtain a material which loses on 5% of its nonlinearity over a time period of 5 years at 80 °C, the glass transition temperature must be in the order of 270 °C [43].

Besides amorphous materials which rely only on their glassy state to freeze the orientation of the chromophores, one can also use modified materials or chromophores which bear cross linking agents for post-poling lattice hardening [39]. For this approach additional chemical bonds are formed after poling in order to lock the chromophores in ideal orientation. Typically, the process is not reversible and offers therefore a high degree of orientational stability. For nonlinear organic materials the crosslinking is often based on Diels-Alder reactions [45-47]. Engineered chromophores allow in-situ cross linking and result in EO material stable up to 200 °C [48].

3.2 State of the art silicon based IQ modulators

3.2.1 IQ modulators based on the plasma dispersion effect

Electro-optic IQ modulators integrated on the silicon platform are key elements to realize optical data links with high data rate, low power consumption and compact footprint. Various concepts emerged to realize phase modulators, the building blocks of an IQ modulator, on the silicon platform. The most common approach to introduce a phase change in a silicon photonic waveguide relies on the modulation of free carriers in the silicon, the plasma dispersion effect, which changes the phase and also the amplitude of the optical signal [49]. They can be grouped in three basic types [7]: The carrier depletion modulator, where a *p-n*-junction is placed within the waveguide and operated under reverse bias. A modulation of the applied voltage leads to a slight modulation of the carriers within the *p-n*-junction and therefore to a modulation of the phase. In a carrier injection modulator a *p-i-n*-junction is placed within the silicon waveguide and it is operated under forward bias. The third variant is the carrier accumulation modulator, similar to the carrier injection modulator but with an insulating layer between the *p* and *n* section within the waveguide.

A sketch of the different modulator types is depicted in Fig. 3.9. For high-speed operation typically carrier depletion modulators are used. The *p-n*-junction is operated in reverse bias to reduce the lifetime of the carriers within the waveguide and achieve high modulation speeds. However, the modulation efficiency of those devices is usually limited to a voltage length product of around 10 Vmm [50]. Devices for practical applications therefore need rather high drive voltages and lengths in the order of several mm. Various demonstrations of depletion type IQ modulators have been done, ranging from QPSK modulation up to 56 GBd [51] over 16QAM modulation at 28 GBd [11] up to 64 QAM modulation at 30 GBd [13]. However, due to the drive voltage of up to 5 V needed for operation, the energy consumption of those modulators is typically around 1 pJ/bit.

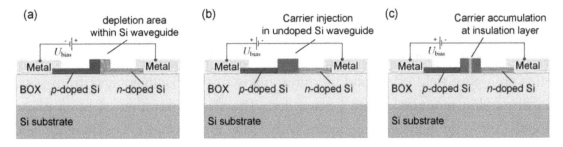

Fig. 3.9: Cross section of different silicon modulator types based on the plasma dispersion effect. (a) Depletion type modulator. A lightly doped *p-n*-junction is placed within the silicon waveguide core, by modulating the reverse bias voltage, the carrier density within the depletion zone can be modulated. (b) Carrier injection modulator. Next to the waveguide core the silicon is *p* and *n*-doped, while the core is intrinsic. Carriers are injected in the waveguide via forward bias. (c) A thin insulation layer is placed within the waveguide core, carriers injected via the doped regions can accumulate in this capacitor structure.

Carrier injection modulators have a higher efficiency and voltage length products well below 1 Vmm but speed is limited by the carrier lifetime within the *p-i-n*-junction [52,53]. Furthermore, the forward bias operation lead to a significant current flow and therefore a high energy consumption.

Carrier accumulation modulators, often realized in a MOS-structure [54], do not feature a static current flow during operation due to the insulating layer. This design can also reach low voltage length products in the order of 1 Vmm and IQ modulators capable of QPSK and 16QAM modulation at a symbol rate of 28 GBd have been demonstrated [12,55]. But still, the energy consumption of these devices is comparatively high. The thin insulation layer leads to a low drive voltage, but also to a large capacitance, which has to be charged and discharged during operation. The associated current flow again increases the energy consumption and the electric drive circuit has to be adapted in order to feed a large capacitance.

Since plasma dispersion modulators all rely on the same physical effect, the energy consumption is basically given by the fact that for a certain phase shift a certain amount of carriers have to be moved in and out of the optical wave guide. The different modulator concepts can therefore only trade drive voltage for drive current, as always roughly the same amount of carriers have to be moved [56].

3.2.2 IQ modulators based on SOH integration

As discussed in Section 3.1, phase modulation in silicon-organic hybrid devices relies on a pure field effect and therefore many of the disadvantages from plasma dispersion modulators can be avoided. The first generation of SOH phase modulators only achieved comparatively low electro-optic coefficients and hence a high voltage length product around 5 Vmm [14,57]. Furthermore, high-speed operation was initially only demonstrated in a single phase modulator [15,58]. More recently, Mach-Zehnder modulators have been demonstrated with high efficiency and very low energy consumption in the fJ/bit range [17,18]. This could be achieved by using a new generation of organic electro-optic materials such as DLD164 and the binary chromophore glass YLD124/PSLD41. However, for those experiments with simple OOK modulation, the modulator was driven without termination to leverage the open circuit voltage of the source and to achieve lowest energy consumption. But when using advanced modulation formats in IQ modulators termination is still needed to maintain a linear frequency response at highest modulation speeds. So far IQ modulators using SOH integration were therefore limited to symbol rates of

28 GBd at 16QAM with an energy consumption of 0.64 pJ/bit and a BER around 1×10^{-3} [16].

Within the work of this thesis, those issues were addressed in order to demonstrate IQ modulators with highest performance and lowest energy consumption. The modulators described in the following are able to generate 16QAM at 28 GBd with an energy consumption two magnitudes lower and at the same time a BER about two magnitudes lower than previous SOH IQ modulators. Furthermore, the symbol rate for 16QAM could be increased up to 40 GBd. This could be achieved mainly by adapting the SOH modulator structure such that it is well matched to a 50 Ohm transmission line and by combining the IQ structure with novel electro-optic organic materials.

3.3 Fabrication of SOH IQ modulators

The devices described within Chapter 3 were fabricated in cooperation with AMO GmbH in Aachen. An SOI wafer with 220 nm crystalline silicon and 3 μm buried oxide is used as substrate. The definition of the silicon waveguide structure is done via E-beam lithography. Hydrogen silsesquioxane (HSQ) is used as negative-tone resist to define the waveguide layer and subsequently the silicon is etched 150 nm. Without removing the first HSQ layer, another layer of HSQ is applied and defined via E-beam lithography to protect the silicon slab from the subsequent 70 nm silicon etch. This ensures a precise definition of the slot waveguide, since the process is self-aligned. After the waveguide definition the HSQ is stripped. To define the areas of low concentration doping in the waveguide region and high concentration doping in the contact areas, optical lithography via an I-line stepper is used. The devices are then *n*-doped with a standard ion implanter and activated through a subsequent annealing step. The aluminum electrodes are deposited via a lift-off process.

Electro-optic organic material is deposited on the SOI structures via spin-coating process under cleanroom conditions according to the corresponding recipe of the material. The remaining solvent in the organic coating is removed by heating the device under vacuum. The devices are then ready for poling and can be directly used in different experiments.

3.4 Generation of advanced modulation formats at low power consumption

The ability to operate an IQ modulator with sub-volt driving voltages has significant advantages compared to the several volts needed in conventional devices. Besides the direct reduction of the power consumption within the modulator, it allows also to omit a RF driver amplifier completely, using only the output stage of the signal generator. This further reduces the power consumption and furthermore, it allows for a higher signal quality, as the noise contribution of the electrical amplifier can be omitted. In the following an SOH modulator is described which enables sub-volt operation at high speed and with high signal quality. This is achieved by exploiting a narrow slot waveguide with an 80 nm slot width, combined with highly efficient electro-optic materials and an impedance matched travelling wave electrode. This section was published in a scientific journal [J7].

[start of paper]

Low-power silicon-organic hybrid (SOH) modulators for advanced modulation formats

Optics Express Vol. 22, Issue 24, pp. 29927-29936 (2014)
DOI: 10.1364/OE.22.029927   © 2014 Optical Society of America

M. Lauermann,[1] R. Palmer,[1,5] S. Koeber,[1,2,6] P. C. Schindler,[1,7] D. Korn,[1] T. Wahlbrink,[3] J. Bolten,[3] M. Waldow,[3] D. L. Elder,[4] L. R. Dalton,[4] J. Leuthold,[1,8] W. Freude,[1,2] and C. Koos[1,2]

[1] *Institute of Photonics and Quantum Electronics, Karlsruhe Institute of Technology, 76131 Karlsruhe, Germany*

[2] *Institute of Microstructure Technology, Karlsruhe Institute of Technology, 76344 Eggenstein-Leopoldshafen, Germany*

[3] *AMO GmbH, 52074 Aachen, Germany*

[4] *University of Washington, Department of Chemistry, Seattle, WA 98195-1700, United States*

[5] *Now with: Coriant GmbH, Munich, Germany*

[6] *Now with: University of Cologne, Chemistry Department, 50939 Köln, Germany*

[7] *Now with: Infinera Corporation, Sunnyvale, CA, USA*

[8] *Now with: Institute of Electromagnetic Fields, Swiss Federal Institute of Technology (ETH), Zurich, Switzerland*

We demonstrate silicon-organic hybrid (SOH) electro-optic modulators that enable quadrature phase-shift keying (QPSK) and 16-state quadrature amplitude modulation (16QAM) with high signal quality and record-low energy consumption. SOH integration combines highly efficient electro-optic organic materials with conventional silicon-on-insulator (SOI) slot waveguides, and allows to overcome the intrinsic limitations of silicon as an optical integration platform. We demonstrate QPSK and 16QAM signaling at symbol rates of 28 GBd with peak-to-peak drive voltages of 0.6 V_{pp}. For the 16QAM experiment at 112 Gbit/s, we measure a bit-error ratio of 5.1×10^{-5} and a record-low energy consumption of only 19 fJ/bit.

1. Introduction

Electro-optical in-phase and quadrature (IQ) modulators are key elements for spectrally efficient coherent transmission in high-speed telecommunication links [59] and optical interconnects [60]. As key requirements, the devices must combine small footprint and high electro-optic bandwidth with low drive voltages and therefore low power consumption. In principle, silicon photonics is a particularly promising platform for realizing such devices, offering high-density photonic-electronic integration at low cost by exploiting large-scale high-yield complementary metal-oxide-semiconductor (CMOS) processing. However, bulk silicon does not feature any second-order optical nonlinearities due to the inversion symmetry of the crystal lattice. Strain can be used to break the crystal symmetry of silicon and to allow for nonzero $\chi^{(2)}$-coefficients [29] which have been exploited to realize Mach-Zehnder modulators [61], but the reported modulation efficiencies are still comparatively low. As a consequence, current silicon-based modulators have to rely on free-carrier depletion or injection in p-n, p-i-n, or in metal-oxide-semiconductor (MOS) structures [7,53,54]. This leads to various trade-offs, particularly when realizing fast and energy-efficient IQ modulators. An all-silicon dual-polarization 16QAM modulator was demonstrated recently [11] using conventional depletion-type p-n-phase shifters. Operating the device with a peak-to-peak drive voltage of 5 V_{pp} at a symbol rate of

28 GBd, an energy consumption of 1.2 pJ/bit and bit error ratios (BER) between 1×10^{-2} and 2.4×10^{-2} have been achieved. These are remarkable results, but the potential for further optimization towards lower drive voltages, reduced energy consumption and smaller footprint seems to be limited by the intrinsically rather low modulation efficiency of depletion-type phase shifters — the voltage-length product of the device [11] amounts to $U_\pi L = 24$ Vmm. Drive voltages can be reduced at the expense of an increased drive current by replacing the p-n-junction with a high-capacitance metal-oxide-semiconductor structure [55]. Such devices require drive voltages of only 1 V_{pp}, but have hitherto been demonstrated for QPSK modulation only. Moreover, both injection and depletion-type modulators suffer from amplitude-phase coupling: Phase modulation in these devices is based on free carriers, the presence of which does not only lead to a change of the refractive index, but also to free-carrier absorption [49]. The associated amplitude modulation may lead to signal distortions when using higher-order modulation formats such as 16QAM, where independent information is encoded on amplitude and phase of the optical carrier.

Silicon-organic hybrid (SOH) integration promises to overcome the inherent limitations of free-carrier plasma dispersion as a modulation mechanism by combining silicon-on-insulator (SOI) slot waveguides with organic electro-optic (EO) claddings [14,32,33]. These slot waveguides can be integrated in photonic-crystal structures that allow reduction of the optical group velocity and hence intensify the interaction of the optical mode with the EO material [58,62,63]. Voltage-length products of $U_\pi L = 0.29$ Vmm were demonstrated by exploiting slow light [64], but practical applications have to cope with the limited bandwidth ranges over which the group delay is flat and dispersion is small, and with the resulting need for wavelength tuning. Non-resonant SOH Mach-Zehnder modulators are more versatile, and modulation experiments with on-off-keying at data rates of 12.5 Gbit/s demonstrated an ultra-low energy consumption of 1.6 fJ/bit [17,65]. At these low data rates the modulator could be driven without termination resistor. However, when using advanced modulation formats two nested Mach-Zehnder modulators are needed, and higher data rates require a termination of the modulator electrodes. As a consequence, the energy consumption increases considerably, e. g., to 640 fJ/bit measured in a 16QAM modulation experiment at a symbol rate of 28 GBd and a BER of 1.2×10^{-3} [16].

In this paper we show that a novel class of organic cladding materials [18,66] can dramatically improve the performance and simultaneously decrease the power consumption of SOH IQ modulators. Using conventional slot waveguides, our devices feature voltage-length products of only $U_\pi L = 0.53$ Vmm — a factor of 20 smaller than that of current all-silicon modulators, and only slightly larger than those of slow-light devices. Operating a 1.5 mm long device at a peak-to-peak drive voltage of only 0.6 V_{pp}, we demonstrate error-free QPSK modulation at a data rate of 56 Gbit/s and 16QAM signaling at 112 Gbit/s with a BER of 5.1×10^{-5} [67]. For the 16QAM experiment, the energy consumption of the modulator is calculated to be only 19 fJ/bit. This is the lowest drive voltage and the lowest energy consumption that have so far been reported for a 28 GBd 16QAM modulator, regardless of the material system. Moreover, the BER of 5.1×10^{-5} corresponds to the lowest value achieved with silicon-based devices for 16QAM at such symbol rates. While this is a first proof-of-concept experiment, we believe that SOH integration has vast potential of further improving the performance of silicon-based IQ modulators, both in terms of symbol rate and modulation efficiency.

2. Silicon-organic hybrid (SOH) modulator

The SOH IQ modulator consists of two nested 1.5 mm long Mach-Zehnder modulators (MZM), each driven by a single coplanar transmission line in ground-signal-ground (GSG) configuration, see Fig. 3.10(a). The phase modulators of the MZM consist of SOI slot waveguides, which are covered by an EO material. When operating the optical slot waveguide in quasi-TE polarization, enhancement of the electric field in the slot leads to strong interaction of the guided light within the EO cladding, see Fig. 3.10(c) for a plot of the E_x component of the optical field. At the same time, the slot waveguide is connected to aluminum transmission lines by thin, lightly *n*-doped silicon slabs, which results in a strong RF modulation field between the rails and hence in a large overlap with the optical mode, see Fig. 3.10(d) [33]. In contrast to conventional silicon modulators based on the plasma dispersion effect, SOH devices do not exhibit amplitude-phase coupling and enable pure phase modulation [15].

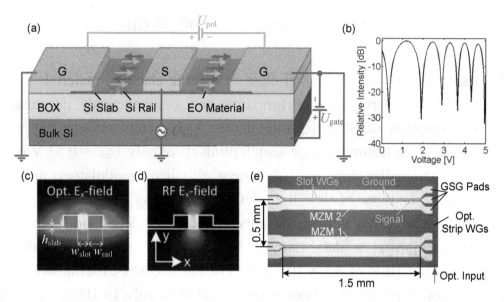

Fig. 3.10: (a) Cross-section of a silicon-organic hybrid (SOH) Mach-Zehnder modulator (MZM). Each arm contains a slot waveguide (rail width w_{rail} = 240 nm, slot width w_{slot} = 80 nm). The device is coated with an electro-optic (EO) organic cladding material, consisting of a mixture of the EO chromophores YLD124 and PSLD41 [18,66]. The waveguides are electrically connected to a ground-signal-ground (GSG) RF transmission line via *n*-doped silicon slabs (thickness h_{slab} = 70 nm). A gate voltage U_{Gate} between the Si substrate and the SOI device layer improves the conductivity of the silicon slab and hence the bandwidth of the device. A poling voltage U_{pol} is initially applied via the floating ground electrodes of the device to align the chromophores in the slots of both waveguides (green arrows). When operating the device via the GSG line, the modulating electric RF field (blue arrows) is anti-parallel (parallel) to the chromophore orientation in the left (right) half of the GSG line. This results in opposite phase shifts in the two arms of the MZM and hence in push-pull operation. (b) Transmission vs. DC voltage of a MZM having 1.5 mm long phase shifters. At bias voltages above 2.9 V, the π-voltage of the device amounts to U_π = 0.35 V. For smaller DC voltages, free charges in the cladding lead to a partial screening of the applied electric field and hence to increased π-voltages. (c) E_x component of the optical field in the slot waveguide. (d) Dominant E_x component of the electrical drive signal. Both fields are confined to the slot, resulting in strong interaction and hence efficient modulation. (e) Optical micrograph of the IQ modulator structure prior to depositing the organic cladding. The GSG transmission lines of both MZM are clearly visible. For contacting, the GSG line is up-tapered at both ends of the slot waveguide section to match the 100 μm pitch of the microwave probe.

The conductivity of the slab is increased by applying a gate voltage U_{gate} between the substrate and the SOI device layer [68], Fig. 3.10(a). The devices were fabricated on an SOI wafer with 220 nm-thick SOI device layer and 3 µm-thick buried oxide (BOX), using a combination of electron-beam lithography for structuring the optical waveguides and optical lithography for the metallization. The fabricated slot width is $w_{\text{slot}} = 80$ nm, and the rail width amounts to $w_{\text{rail}} = 240$ nm.

The transmission line consists of 200 nm thick aluminum strips in a ground-signal-ground (GSG) configuration [38]. The parts of the silicon slab which are overlapping with the aluminum are highly n-doped to achieve an Ohmic contact. The central conductor of the GSG transmission line has a width of approximately 10 µm, and the signal-ground electrode separation is approximately 5 µm. The transmission line is designed with CST Microwave Studio, the model includes the doping of the silicon slabs by introducing a small conductivity in the material. Electrical characterization of the fabricated device with a 50 Ω probe shows low back reflections of the RF signal with S_{11} magnitudes of less than −20 dB at frequencies of up to 40 GHz and the actual device impedance amounts to 50 ± 5 Ω. The group refractive index mismatch is $n_{\text{opt}} - n_{\text{RF}} = 0.4$ and thus the walk-off is so small that it does not significantly influence the bandwidth at a device length of 1.5 mm.

For the EO material, we use a mixture of the multi-chromophore dendritic molecule PSLD41 (75 wt.-%) and the chromophore YLD124 (25 wt.-%) [18,66]. This binary-chromophore organic glass (BCOG) has a refractive index of $n = 1.73$ and shows in bulk material an r_{33} coefficient of 285 pm/V [66]. The material-related in-device r_{33} of 230 pm/V reported in [18] with this BCOG corresponds to the highest value shown so far for SOH devices. The material is deposited via spin coating, and a poling step is performed for aligning the randomly oriented chromophores. To this end, the EO material is heated close to its glass transition temperature of 97 °C while applying a poling voltage across the two floating ground electrodes, see Fig. 3.10(a). This leads to an ordered orientation of the dipolar chromophores along the poling field, indicated by green arrows in Fig. 3.10(a). The orientation of the chromophores is frozen by cooling the device back to room temperature before removing the poling voltage. For operating the device, an RF signal is coupled to the GSG transmission line, leading to a modulating electric field as indicated by blue arrows in Fig. 3.10(a). The RF field and the chromophore orientation are antiparallel to each other in the left half of the GSG transmission line, and parallel in the right half. This leads to phase shifts of opposite signs in the two arms of the MZM as needed for push-pull operation. To establish good electrical contact for the microwave probes, the EO material is mechanically removed at the contact pads. An intentional imbalance in the parent MZM allows to adjust a π/2 phase shift between the in-phase (I) and quadrature (Q) component via wavelength tuning, see Fig. 3.11(a). Characterization of one of the MZM at DC voltages exhibits a π-voltage U_π for the push-pull modulator of only 0.35 V for bias voltages of more than 2.9 V, see Fig. 3.10(b). The phase shifters of the device are 1.5 mm long, resulting in a voltage-length product as low as $U_\pi L = 0.53$ Vmm for push-pull operation, which enables low-power operation of the IQ modulator. For smaller bias voltages, we observe increased spacings of the transmission dips in Fig. 3.10(b) and hence increased π-voltages. This is attributed to free charges in the cladding that lead to a partial screening of the applied fields at small bias voltages. Due to the low mobility of the free charges, this is only observable for low frequencies and does not impede high frequency operation. An optical micrograph of the SOH device prior to depositing the organic cladding is shown in Fig. 3.10(e). From the measured U_π we can calculate the electro-optic coefficient r_{33} with the relation [18,38]

$$r_{33} = \frac{w_{slot}\lambda_c}{2LU_\pi \Gamma n_{slot}^3} \quad (3.32)$$

where w_{slot} denotes the slot width, λ_c is the wavelength of the optical carrier, L the modulator length and n_{slot} the refractive index of the organic material in the slot. The field interaction Γ is linked to the fraction of the optical power that interacts with the modulation field [18,38]. Our device geometry is optimized for a maximum field interaction, which amounts to $\Gamma = 0.22$. Using Eq. (3.32) the electro-optic coefficient is determined to be 104 pm/V. The poling efficiency r_{33}/E_{poling} amounts to 0.28 nm^2/V^2 for this particular device. For practical applications of SOH devices, the long-term stability of the organic cladding is of high importance. Materials have become available featuring glass transition temperatures of more than 130 °C while maintaining electro-optic coefficients in excess of 100 pm/V [47]. The investigation of aging and temperature stability of organic EO materials is subject to ongoing research. It can be expected that the stability of the materials can be further improved by synthetically modified chromophores that bear specific crosslinking agents for post-poling lattice hardening or by increasing the molar mass of the chromophores. The viability of the first approach has been demonstrated for similar EO compounds [46,69] where material stability of up to 250 °C has been achieved.

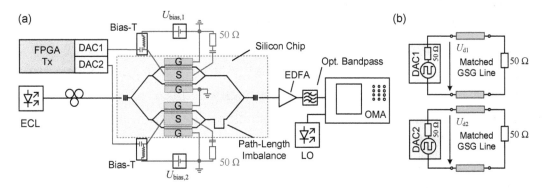

Fig. 3.11: Schematic of the experimental setup. (a) Two nested MZM form an IQ modulator. An intentional path-length imbalance in the parent Mach-Zehnder interferometer allows for adjusting the phase difference of in-phase (I) and quadrature-phase (Q) component of the signal to $\pi/2$ by wavelength tuning. Electrical multilevel drive signals are generated by field-programmable gate arrays (FPGA) and high-speed digital-to-analog converters (DAC), the outputs of which are directly coupled to the silicon chip via microwave probes. Bias-Ts are used to apply DC bias voltages to the MZM for adjusting the operating points. An external-cavity laser (ECL) is used as an optical source and coupled to the chip via fibers and grating couplers. The optical output signal is amplified by an EDFA and subsequently fed into an optical modulation analyzer (OMA) with a second laser as local oscillator (LO). (b) Equivalent-circuit diagram of the experimental setup for calculation of the energy consumption. Each MZM features a 50 Ω GSG transmission line which is terminated by an external 50 Ω resistor and is driven directly by one DAC, represented by an ideal voltage source and an internal resistance of 50 Ω. The GSG line of the MZM is matched to the 50 Ω output of the DAC. To estimate the energy consumption, we can replace the transmission line and its termination by an equivalent resistor of $R = 50$ Ω.

3. Signal generation experiment

The experimental setup for generation of data signals is depicted in Fig. 3.11(a). Light from an external-cavity laser is coupled to the chip via grating couplers.

The electrical non-return-to-zero (NRZ) drive signals are generated in software-defined multi-format transmitters, based on field-programmable gate arrays (FPGA) [70] and operated at a symbol rate of 28 GBd. The transmitters comprise a pair of digital-to-analog converters (DAC) that are directly coupled to the two on-chip MZM using RF probes, without any additional amplification. The electrical transmission line of each MZM features an impedance close to 50 Ω, matched to the output impedance of the DAC, and is terminated with an external 50 Ω resistor to avoid back reflections at the end of the transmission line. A gate field of E_{Gate} = 0.1 V/nm is applied between the silicon substrate and the SOI device layer to increase the bandwidth [68]. To apply the gate voltage, the SOH chip is placed on an electrically isolated sample mount which can be set to a defined potential with respect to the ground of the microwave probes. To avoid high gate voltages across the 3 μm BOX in the future, we plan to deposit a silicon layer on top of the silicon slabs, isolated from the slabs by a thin silicon oxide film [15]. This structure features a thin gate oxide with only a few nanometers thickness and can thereby reduce the necessary gate voltage to a few volts only. Negligible current flow is associated with the applied gate voltage, and the energy consumption of the device is hence not increased.

The fiber-to-fiber insertion loss of the silicon chip, obtained from a transmission measurement without modulation signal or gate voltage and with the modulator tuned to maximum transmission, amounts to 27 dB. It is compensated by an erbium-doped fiber amplifier (EDFA). The rather high optical insertion loss of our modulator originates mainly from the losses of the non-optimized grating couplers (together 10 dB), the losses of the 9 mm long access strip waveguides (0.5 dB/mm due to sidewall roughness), from the loss of the optimized strip-to-slot converters [71] and multimode interference couplers (MMI) (together 1 dB), and from the propagation loss in the 1.5 mm long slot waveguide section of the phase shifter (approx. 6.5 dB/mm caused by roughness and below 1 dB/mm by doping [49]). A bandpass-filter is inserted after the EDFA to remove out-of-band amplified spontaneous emission (ASE) noise from the EDFA. The signal is then detected by an optical modulation analyzer (OMA), using a second external-cavity laser as a local oscillator (LO) in an intradyne configuration [72]. Standard digital post-processing comprising polarization demultiplexing, compensation of the frequency offset between the transmitter laser and the LO, phase recovery and channel equalization is applied at the OMA. The optical insertion loss of the current modulator is still rather high, but we expect that losses can be reduced considerably in future device generations. As an example, the insertion loss of undoped slot waveguides can be decreased well below 1 dB/mm by using asymmetric waveguide geometries and by improving the sidewall roughness [73]. Moreover, optimization of doping profiles outside the region where the light is guided might allow to operate the device without a gate field which causes 1 dB – 2 dB optical loss, and to decrease the total carrier-induced propagation loss that currently amounts to approx. 3 dB/mm. Similarly, using optimized grating couplers [74] or photonic wire bonds [75,76], the coupling loss to external fibers can be improved to below 2 dB per chip facet.

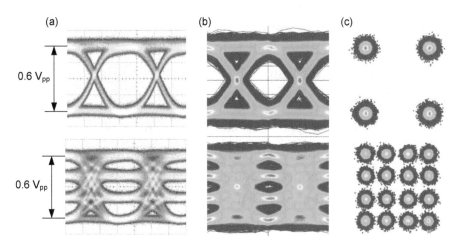

Fig. 3.12: Measured eye diagrams and constellations for a symbol rate of 28 GBd. (a) Eye diagram of the electrical drive signals for one MZM in the case of QPSK (top) and 16QAM (bottom). Peak-to-peak drive voltages of 0.6 V_{pp} are sufficient to generate a high-quality optical signal. The eye diagrams were obtained by connecting an oscilloscope with a 50 Ω input impedance to the DAC that are used to drive the modulators. (b) Corresponding optical eye diagrams, measured with the optical modulation analyzer. (c) Corresponding constellation diagrams of the optical signals. The measured EVM_m of the QPSK signal is 10.9%; no errors could be measured. From the measured EVM_m, the BER can be estimated to be well below to 1×10^{-10}. For the 16QAM signal, the EVM_m is 8.7% and the measured BER amounts to 5.1×10^{-5}.

In a first modulation experiment, a QPSK signal at 28 GBd is generated, derived from a pseudorandom binary sequence (PRBS) with a length of $2^{11} - 1$.

The electrical eye diagram of one channel is depicted in Fig. 3.12(a). The eye diagram was obtained by connecting an oscilloscope with 50 Ω input impedance to the output of the DAC, and the measured voltage levels are hence identical to the drive voltages of the 50 Ω MZM. The peak-to-peak voltages of the electrical drive signals are found to be 0.60 V for the I-channel and 0.63 V for the Q-channel. The voltages are measured at the end of the RF cable, before connecting to the microwave probes. The measured optical eye diagram of one channel and the constellation diagram are depicted in Fig. 3.12(b) and (c), top row. As a quantitative measure of the signal quality, we use the error vector magnitude (EVM_m), which describes the effective distance of a received complex symbol from its ideal position in the constellation diagram, using the maximum length of an ideal constellation vector for normalization. For the QPSK signal, the EVM_m is measured to be 10.9 %. If we assume additive white Gaussian noise (AWGN) as main limitation of our received signal, we can relate this EVM_m to a bit error ratio (BER) well below 1×10^{-10} [77]. Within the recorded 62.5 μs signal section, we measure 1.7×10^6 symbols, so a direct measurement of a BER was not possible.

In a second experiment, a 16QAM signal is generated [67]. The electrical 4-level signal of the I-channel is shown in Fig. 3.12(a), along with the optical eye diagram and the corresponding constellation diagram measured at 28 GBd, Fig. 3.12(b) and (c), bottom row. The measured BER is 5.1×10^{-5} and the EVM_m of the signal is 8.7%. This EVM_m would correspond to a calculated BER of 4.8×10^{-5}, which is in very good agreement with the measured value. The good match between the calculated and measured BER supports the assumption that the signal quality is limited by AWGN only. The signal to noise ratio (SNR) is measured at the OMA after coherent reception and amounts to 20.4 dB within the evaluated signal bandwidth of 40 GHz. This corresponds to an optical SNR (OSNR) of 22.4 dB at a reference bandwidth of 0.1 nm with noise in both polarizations [10]. Furthermore, we see in Fig. 3.12(c) very regular constellation diagrams without distortion or rotation, from which we can conclude that the SOH modulator provides indeed pure phase modulation without any residual amplitude-phase coupling. For high-speed operation, the π-voltage increases due to bandwidth limitations [38]. A characterization of one MZM results in a 6 dB electro-optic-electric (3 dB electro-optic) bandwidth of 18 GHz and U_π at symbol rates of 28 GBd can be estimated to 0.7 V, resulting in a corresponding voltage-length product of $U_\pi L = 1.05$ Vmm and a modulation index of $U/U_\pi = 0.86$.

4. Energy consumption

For estimating the energy consumption of our devices, we use an equivalent circuit of the MZM and the driver electronics, see Fig. 3.11(b): Each DAC, represented by an ideal voltage source and an internal resistance of 50 Ω, is connected to a 50 Ω transmission line that represents the MZM and that is terminated by a matched 50 Ω resistor. For estimating the power consumption of the MZM, we can replace the transmission line and its termination by an equivalent resistor of $R = 50$ Ω. The power consumption of a single MZM is then given by the power dissipation in this resistor, and the energy per bit for the IQ modulator is obtained by adding the power consumptions of the two MZM and dividing by the total data rate. For QPSK, assuming rectangular non-return-to-zero drive signals and equal probability of all symbol states, the energy consumption per bit for the IQ modulator can be calculated according to

$$W_{\text{bit,QPSK}} = \left[\left(\frac{U_{d1}}{2} \right)^2 \frac{1}{R} + \left(\frac{U_{d2}}{2} \right)^2 \frac{1}{R} \right] \times \frac{1}{r_{\text{QPSK}}}. \quad (3.33)$$

The quantities $U_{d1,2}$ are the peak-to-peak voltages of the electrical drive signals, measured across the equivalent resistor of $R = 50$ Ω, which amount to 0.60 V for the I and 0.63 V for the Q-channel. The aggregate data rate is $r_{\text{QPSK}} = 56$ Gbit/s for the QPSK signal, leading to a power consumption of 68 fJ/bit.

For 16QAM, the energy consumption per bit can be calculated similarly. For each MZM, the power of the drive signal is dissipated in the equivalent 50 Ω-resistor. However, the drive signal of each MZM now consists of two amplitude levels, which differ by a factor of 1/3. Assuming again equal distribution of the various symbols, the energy of each MZM can be calculated by averaging over the dissipated energy of both amplitude levels. The energy per bit is then obtained by adding the power consumption of the two MZM and dividing by the total data rate,

$$W_{\text{bit},16\text{QAM}} = \left(\begin{array}{c} \dfrac{1}{2}\left[\left(\dfrac{U_{d1}}{2}\right)^2 \dfrac{1}{R} + \left(\dfrac{1}{3}\dfrac{U_{d1}}{2}\right)^2 \dfrac{1}{R}\right] \\ + \dfrac{1}{2}\left[\left(\dfrac{U_{d2}}{2}\right)^2 \dfrac{1}{R} + \left(\dfrac{1}{3}\dfrac{U_{d2}}{2}\right)^2 \dfrac{1}{R}\right] \end{array} \right) \times \dfrac{1}{r_{16\text{QAM}}}. \qquad (3.34)$$

Using the measured peak-to-peak voltages of $U_{d1} = 0.60$ V and $U_{d2} = 0.63$ V and a data rate of $r_{16\text{QAM}} = 112$ Gbit/s, the energy consumption of the modulator is found to be 19 fJ/bit for the case of 16QAM signaling. This is, to the best of our knowledge, the lowest energy consumption that has so far been reported for a 16QAM modulator at such symbol rates, regardless of the material system. Applying DC bias and gate voltages to the modulator leads to a very small current flow around 1 nA. The corresponding energy consumption amounts to few aJ/bit and can safely be neglected. It should be noted that the power consumption has been estimated for a modulator with an ideal impedance of 50 Ω. Input impedances deviating from this value would lead to non-perfect impedance matching in our experiment and hence to partial reflection of the RF drive signal. As a consequence, the power consumption of the real device should then even be slightly smaller than the values estimated above, and our estimation of the power consumption hence represents a worst-case scenario.

5. Summary

We experimentally show that the combination of novel EO materials and slot waveguide structures enables silicon-based IQ modulators with unprecedented performance. We demonstrate generation of 28 GBd QPSK and 16QAM signals, leading to data rates of 56 Gbit/s, and 112 Gbit/s, respectively. Using record-low peak-to-peak drive voltages of $U_d = 0.6$ V, we obtain error-free QPSK modulation and 16QAM signals with a BER of 5.1×10^{-5}. The drive signal is derived directly from the output of the DAC, without the need for an additional driver amplifier. For the 16 QAM experiment, we estimate an energy consumption of only 19 fJ/bit — the lowest value reported so far for 16QAM signaling at 28 GBd.

[end of paper]

3.5 Generation of data signals with high symbol rates

To explore the limits of the performance for such devices, the signal quality of SOH modulators was evaluated at highest modulation speeds. This section was published in a scientific journal [J9].

[start of paper]

40 GBd 16QAM Signaling at 160 Gbit/s in a Silicon-Organic Hybrid (SOH) Modulator

IEEE Journal of Lightwave Technology, Vol. 33, Issue 6, pp. 1210–1216, (2015)
DOI: 10.1109/JLT.2015.2394211 © 2015 IEEE

M. Lauermann,[1] S. Wolf,[1] P. C. Schindler,[1,2] R. Palmer,[1,3]
S. Koeber,[1,4,5] D. Korn[1], L. Alloatti,[1,6] T. Wahlbrink,[7] J. Bolten,[7]
M. Waldow,[7] M. Koenigsmann,[8] M. Kohler,[8] D. Malsam,[8]
D. L. Elder,[9] P. V. Johnston,[9] N. Phillips-Sylvain,[9] P. A. Sullivan,[10]
L. R. Dalton,[9] J. Leuthold,[1,11] W. Freude,[1,4] C. Koos,[1,4]

[1] *Karlsruhe Institute of Technology, Institute of Photonics and Quantum Electronics (IPQ), 76131 Karlsruhe, Germany*

[2] *now with: Infinera Corporation, Sunnyvale, CA, USA*

[3] *now with: Coriant GmbH, Munich, Germany*

[4] *Karlsruhe Institute of Technology, Institute of Microstructure Technology (IMT), 76344 Eggenstein-Leopoldshafen, Germany*

[5] *now with: University of Cologne, Chemistry Department, 50939 Köln, Germany*

[6] *Now with: Massachusetts Institute of Technology, Research Lab of Electronics (RLE), Cambridge, MA 02139, United States*

[7] *AMO GmbH, 52074 Aachen, Germany*

[8] *Agilent Technologies, 71034 Boeblingen, Germany*

[9] *University of Washington, Department of Chemistry, Seattle, WA 98195-1700, United States*

[10] *Montana State University, Department of Chemistry & Biochemistry, Bozeman, MT 59717, United States*

[11] *Now with: Institute of Electromagnetic Fields, ETH Zurich, Switzerland*

3 Energy efficient high-speed IQ modulators

We demonstrate for the first time generation of 16-state quadrature amplitude modulation (16QAM) signals at a symbol rate of 40 GBd using silicon-based modulators. Our devices exploit silicon-organic hybrid (SOH) integration, which combines silicon-on-insulator slot waveguides with electro-optic cladding materials to realize highly efficient phase shifters. The devices enable 16QAM signaling and quadrature phase shift keying (QPSK) at symbol rates of 40 GBd and 45 GBd, respectively, leading to line rates of up to 160 Gbit/s on a single wavelength and in a single polarization. This is the highest value demonstrated by a silicon-based device up to now. The energy consumption for 16QAM signaling amounts to less than 120 fJ/bit – one order of magnitude below that of conventional silicon photonic 16QAM modulators.

1. Introduction

Fast and efficient in-phase/quadrature-phase (IQ) modulators are key elements for high-speed links in telecom and datacom networks [78]. To maximize the data rate that can be transmitted on a single wavelength channel, both large symbol rates and the ability to use higher-order modulation formats are essential [79]. At the same time, minimizing the power consumption of the devices is of utmost importance regarding high-density integration and scalability of interconnect counts.

Silicon photonics is a particularly attractive platform for realizing electro-optic modulators, leveraging mature complementary metal-oxide-semiconductor (CMOS) processing and enabling high-density integration of photonic devices along with electronic circuitry. However, the inversion symmetry of the silicon crystal lattice inhibits electro-optic effects, thereby making high-performance IQ modulators challenging. As a consequence, conventional silicon modulators have to rely on carrier depletion or carrier injection in p-n, p-i-n or metal-oxide-semiconductor (MOS) structures [7,53,54,80]. Using a depletion-type device, generation of quadrature phase shift keying (QPSK) signals was recently shown at a symbol rate of 56 GBd resulting in a total line rate of 112 Gbit/s [81]. However, when using more advanced modulation formats, the achievable symbol rates are still significantly lower – record values amount to 28 GBd demonstrated for dual-polarization 16-state quadrature amplitude modulation (16QAM) [11], which leads to a line rate (net data rate) of 112 Gbit/s

3.5 Generation of data signals with high symbol rates

(93.3 Gbit/s) encoded on each polarization. The performance of these devices is inherently limited by the underlying depletion-type phase shifters, which exhibit rather low efficiencies with typical voltage-length products $U_\pi L$ of 10 Vmm or more. As a consequence, large drive voltages, on the order of several volts, have to be used, leading to high energy consumption – for 16QAM modulation at 28 GBd, the modulation energy amounts to approximately 1.2 pJ/bit at a peak-to-peak drive voltage of 5 V_{pp} [11]. In addition, phase modulation based on the plasma dispersion effect is inevitably linked to amplitude modulation due to free-carrier absorption. This may eventually hamper the generation of advanced modulation formats with high order, where phase and amplitude of the signals have to be controlled independently of one another.

In this paper we show that silicon-organic hybrid (SOH) integration can overcome these limitations. We use silicon-on-insulator (SOI) slot waveguides [82] and combine them with electro-optic (EO) cladding materials to realize Pockels-type phase shifters [14,15,32,33]. SOH integration enables remarkably small voltage-length products of down to 0.5 Vmm measured at DC [18], and at the same time avoids unwanted amplitude-phase coupling and thereby enables higher-order modulation formats [16,83]. We demonstrate 16QAM and QPSK signaling at symbol rates of 40 GBd and 45 GBd, respectively, leading to line rates (net data rates) of up to 160 Gbit/s (133.3 Gbit/s) on a single polarization [84]. This is the highest value achieved by a silicon-based modulator up to now. The energy consumption of our 16QAM device is estimated to be 120 fJ/bit at 40 GBd – one order of magnitude better than for best-in-class 28 GBd 16QAM all-silicon modulators. The work builds upon and expands our earlier experiments, where we have demonstrated energy-efficient on-off-keying [85], generation of multi-level amplitude modulation at symbol rates of up to 84 Gbit/s [83] and 16QAM modulation at 28 GBd [67].

2. Silicon-Organic Hybrid Modulator

SOH modulators exploit interaction of the guided light with the electro-optic cladding material under the influence of a modulating RF field. A cross section of an SOH Mach-Zehnder modulator (MZM) is depicted in Fig. 3.13(a). Each phase shifter consists of a silicon slot waveguide which is covered by the organic EO material. Fig. 3.13(b) and Fig. 3.13(c) show the optical E_x field. The discontinuity and the high refractive index contrast at the silicon – slot interface leads to a field enhancement within the slot and therefore to a large interaction of the light with the organic EO material [33]. The rails of the slot waveguide have a width of w_{rail} = 240 nm and are connected to a coplanar ground-signal-ground (GSG) transmission line via thin (h_{slab} = 70 nm), slightly conductive, n-doped silicon slabs. A voltage applied to the transmission line will drop predominantly across the w_{slot} = 80 nm wide slot, thereby generating a large electric field. A plot of the E_x component of the electrical field can be seen in Fig. 3.13(d). This configuration ensures excellent overlap between the modulation field and the optical fields, leading to high modulation efficiency. The RF transmission line comprises the metal traces and the silicon slot waveguides and is designed for a wave impedance of 50 Ω. This is confirmed by measurements – at RF signals of up to 40 GHz, we find wave impedances of 50 ± 5 Ω.

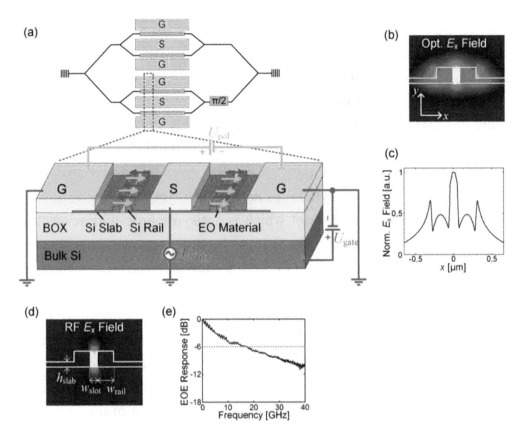

Fig. 3.13: Silicon-organic hybrid (SOH) modulator. (a). Schematic of the IQ modulator and cross-section of a single SOH Mach-Zehnder Modulator (MZM). The slot waveguides have a rail width of w_{rail} = 240 nm and a slot width of w_{slot} = 80 nm. As an EO cladding, a mixture of the chromophores YLD124 (25 wt.%) and PSLD41 (75 wt.%) is deposited via spin coating. Thin *n*-doped silicon slabs with (h_{slab} = 70 nm) are used to electrically connect the rails to the metal strips of an RF transmission line in a ground-signal-ground configuration. A poling process is used to align the chromophores in both waveguides along the same direction. Operating the device via the GSG transmission line results in opposite phase shifts in the two arms of the MZM (push-pull operation). The π-voltage at DC is U_π = 0.9 V. A gate voltage U_{Gate} between the Si substrate and the SOI device layer improves the conductivity of the silicon slab, resulting in an electro-optic bandwidth of the device of 18 GHz. (b) Contour plot of the normalized E_x component of the optical field in the slot waveguide. (c) Plot of the E_x-component of the optical mode field as a function of the horizontal position x at half the waveguide height (y = 110 nm). Discontinuities of the E_x-component at the slot sidewalls lead to strong field enhancement in the slot. (d) E_x component of the electrical RF drive signal below the *RC* limit. The silicon slabs are doped such that the applied RF voltage drops predominantly across the slot. As a consequence, the RF mode and the optical mode are both well confined to the slot, resulting in strong interaction and hence in an efficient modulation. (e) Measured frequency response of a 1.5 mm-long MZM. The 6 dB electrical-optical-electrical (3 dB electro-optic) bandwidth amounts to 18 GHz.

The device was fabricated on an SOI wafer with a 220 nm-thick device layer and a 3 μm-thick buried oxide (BOX) using electron beam lithography for defining the silicon waveguides and optical lithography for the metallization. The chip is coated with a mixture of the electro-optic multi-chromophore dendritic molecule PSLD41 (75 wt.%) and the chromophore YLD124 (25 wt.%) [18,66]. The cladding material is poled by heating it close to its glass transition temperature while applying a poling voltage across the two floating ground electrodes of each MZM. Half of the voltage drops across each slot, resulting in an orientation of the chromophores in the slot which is antisymmetric with respect to the signal electrode, as indicated in Fig. 3.13(a) by green arrows. The blue arrows indicate the RF field applied to the GSG electrodes after poling, which results in opposite phase shifts in the two arms of the MZM and enables push-pull operation.For DC fields, the π-voltage of one push-pull MZM amounts to 0.9 V. Taking into account the device length of 1.5 mm, this corresponds to a voltage-length product of $U_\pi L$ = 1.35 Vmm. This is higher than previously published values of SOH devices [14] and is attributed to different device geometry and the resulting difference in poling efficiency. The bandwidth of the SOH devices is dictated by the RC time constant of the slot waveguide: The slot corresponds to a capacitor which is charged and de-charged via the resistive silicon slabs. To increase the conductivity of the slab by a charge accumulation layer and hence to increase the bandwidth, a static gate voltage U_{gate} is applied between the substrate and the top silicon layer [68], Fig. 3.13(a). A bandwidth measurement of the current MZM results in a 6 dB electrical-optical-electrical (3 dB electro-optic) bandwidth of 18 GHz, see Fig. 3.13(e), with significant potential for further improvement [86]. The roll-off is relatively smooth and resembles that of an RC low-pass, thereby still allowing 40 GBd 16QAM and 45 GBd QPSK signaling using root-raised-cosine Nyquist pulses.

For practical applications of SOH devices, the long-term stability of the organic cladding is of high importance. Recently novel materials have become available, featuring glass transition temperatures of more than 130 °C while maintaining electro-optic coefficients in excess of 100 pm/V [47]. The investigation of aging and temperature stability of organic EO materials is subject to ongoing research. It can be expected, that the stability of the materials can be further improved by synthetically modified chromophores that bear specific crosslinking agents for post-poling lattice hardening or by increasing the molar mass of the chromophores. The viability of the first approach has been demonstrated for similar EO compounds [46,69] where material stability of up to 250 °C has been achieved.

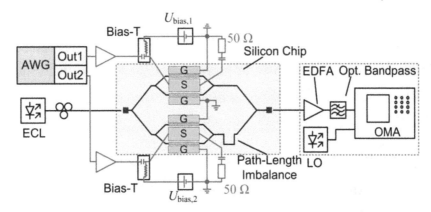

Fig. 3.14: Schematic of the experimental setup. Two nested MZM form an IQ modulator. An intentional imbalance allows for adjusting the π/2 phase offset of the in-phase (I) and the quadrature-phase (Q) component by wavelength tuning. Electrical multilevel drive signals are generated by a Keysight M8195A arbitrary waveform generator (AWG) operating at 65 GSa/s, the outputs of which are amplified and coupled to the silicon chip via RF probes. Bias-Ts are used to connect a DC voltage for controlling the operation point. The optical output is amplified by an EDFA, bandpass filtered and subsequently fed into an optical modulation analyzer (OMA) for intradyne detection. Standard digital post processing is performed at the OMA for equalization.

3. Experiment

The setup for the modulation experiments is depicted Fig. 3.14. Two MZM are nested in an on-chip IQ configuration. The parent Mach-Zehnder interferometer features a path length imbalance, which is used to adjust the π/2 phase shift between the I and the Q signal via wavelength tuning. The drive signals for QPSK and 16QAM modulation are generated by a Keysight M8195A arbitrary waveform generator (AWG) operated at 65 GSa/s. We use root-raised-cosine pulses with a roll-off factor of $\beta = 0.35$ to reduce the occupied spectral bandwidth and to exploit the benefits of a matched filter at the receiver. The transmitted data are derived from a pseudo-random bit sequence with a length of $2^{11}-1$. After the AWG, the analog signals are amplified to peak-to-peak voltages of approximately 1.8 V_{pp} by two linear RF-amplifiers, which drive the GSG-electrodes of the on-chip IQ-modulator via microwave probes. A 50 Ω termination is used at the end of each transmission line to avoid back-reflection of the RF signal. For each MZM a DC voltage is applied to the device via bias-Ts to set the operating point of the modulator. To improve the bandwidth of the device, a static gate field of $E_{Gate} = 0.1$ V/nm is applied between the silicon substrate and the device layer. Grating couplers are used to couple laser light at $\lambda = 1550$ nm from an external cavity laser (ECL) to the silicon waveguide. After the device, the light is amplified by an erbium doped fiber amplifier (EDFA), followed by an optical band-pass filter (2 nm passband) to suppress out-of-band amplified spontaneous emission (ASE). An optical modulation analyzer (OMA) with two real-time oscilloscopes (80 GSa/s) serves as a receiver. A second ECL is used as a local oscillator (LO) for intradyne reception. Digital post-processing comprising polarization demultiplexing, phase recovery, compensation of the frequency offset between signal and LO, and channel equalization is performed by the OMA. The insertion loss of the silicon chip amounts to 27 dB, where 10 dB – 12 dB are caused by fiber-chip coupling losses. This rather high loss can be significantly reduced in future devices: By employing optimized grating coupler [74] or photonic wirebonds [76,87] the coupling loss to the fibers can be reduced from 10 dB to less than 4 dB. The current losses of 5 dB in the 9 mm long access strip waveguide can be reduced to below 1 dB by improving the sidewall roughness and reducing the length. The strip-to-slot converters and multimode interference couplers (MMI) are already optimized and contribute only 1 dB to the total loss. To reduce the 11 dB losses of the 1.5 mm long slot waveguide to below 3 dB, asymmetric slot waveguide geometries can be used [73] together with an optimization of the doping profile of the phase shifter sections. We estimate that these measures will permit reducing the total on-chip excess loss of the device to less than 5 dB and the fiber-fiber insertion loss to less than 9 dB.

3.5 Generation of data signals with high symbol rates

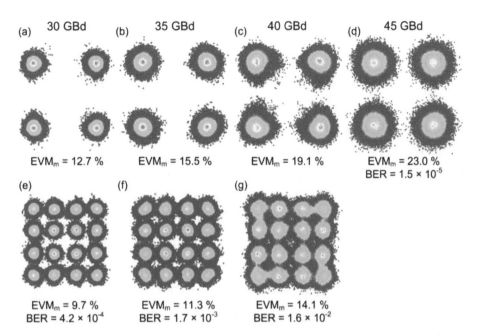

Fig. 3.15: Optical constellation diagrams. (a) – (d): QPSK signals at 30 GBd, 35 GBd, 40 GBd, and 45 GBd. No bit errors were detected within our record length of 62.5 µs for symbol rates of 30 GBd and 35 GBd, and the error vector magnitude (EVM$_m$) indicate error free signals with BER < 10^{-9}. QPSK signals at 40 GBd and 45 GBd are well below the threshold for hard-decision FEC with 7 % overhead. At 40 GBd the number of measured errors was not sufficient for a reliable BER estimation, at 45 GBd the measured BER amounts to 1.5×10^{-5}. (e) – (g): 16 QAM signals for 30 GBd, 35 GBd, and 40 GBd. At data rates of up to 35 GBd the measured BER is below the hard-decision FEC threshold. For 40 GBd 16QAM signals, the BER is below the threshold for soft-decision FEC with 20 % overhead.

For testing the performance of the devices, we use QPSK signals up to a symbol rate of 45 GBd and 16QAM signals up to 40 GBd. Measurement of the bit error ratio (BER) allows for direct assessment of the signal quality. However, within our memory-limited record length of 62.5µs, we can only measure a maximum of 2.8×10^6 symbols, which does not allow a measurement of BER smaller than 1×10^{-6}. As a complementary measure of the signal quality, we therefore use the error vector magnitude (EVM$_m$), which describes the effective distance of a received complex symbol from its ideal position in the constellation diagram, using the maximum length of an ideal constellation vector for normalization. The EVM$_m$ can be directly translated into a BER assuming the signal is impaired by additive white Gaussian noise only [77]. In our experiment we find a very good agreement between BER values calculated from the EVM$_m$ and the measured BER, supporting our assumption that the channel is limited by Gaussian noise. Fig. 3.15 shows constellation diagrams for the various generated signals. In Fig. 3.15(a)-(d), received QPSK constellation diagrams at data rates between 30 GBd and 45 GBd are depicted. At 45 GBd the EVM_m amounts to 23 % and the measured BER is 1.5×10^{-5}, well below the limit of 4.5×10^{-3} for second-generation hard-decision forward error correction (FEC) with 7 % overhead [88]. At 40 GBd the number of measured errors was not sufficient for a reliable BER estimation. At symbol rates of 35 GBd and below, the QPSK signals can be considered error free: No errors were measureable, and the EVM_m corresponds to a BER well below 1×10^{-9}. The constellation diagrams for 16QAM are depicted in Fig 3(e) - (g). 16QAM signaling is demonstrated up to a symbol rate of 40 GBd. At symbol rates of up to 35 GBd, the measured BER is below the threshold for hard-decision FEC. For 40 GBd, the BER increases significantly, but is still below the threshold of 2.4×10^{-2} for third-generation soft-decision FEC with 20 % overhead [89].

The results of the signal generation experiments are summarized in Fig. 3.16, where the EVM_m is plotted for different symbol rates. The horizontal dashed lines in Fig. 3.16 indicate the calculated EVM_m [77] of the various BER threshold levels. For QPSK signals in Fig. 3.16(a), the dashed line corresponds to the BER of 1×10^{-9}, below that the signals can be considered error free. This applies to the signals measured at 30 GBd and 35 GBd. All other measured QPSK signals are well below the hard-decision FEC limit which corresponds to an EVM_m of 38.3 % and is outside the scale of the vertical axis. For 16QAM signals in Fig. 3.16(b), the horizontal lines indicate the EVM_m corresponding to the BER thresholds for hard-decision and soft-decision FEC, requiring 7 % and 20 % overhead, respectively. For symbol rates up to 35 GBd the EVM_m is below the threshold for hard-decision FEC; for 40 GBd it is still below the soft-decision threshold, consistent with the directly measured BER indicated in Fig. 3.15(e)-(g). Using 35 GBd 16QAM signals with hard-decision FEC, the line rate amounts to 140 Gbit/s, and the net data rate amounts to 130.8 Gbit/s. For 16QAM at 40 GBd, the line rate is 160 Gbit/s – the highest value hitherto achieved by a silicon photonic modulator on a single polarization with measured BER figures comparable to those achieved in reference experiments [11]. Taking into account the 20 % overhead for soft-decision FEC coding, the net data rate amounts to the record value of 133.3 Gbit/s. The signal quality in the 40 GBd experiment is limited by the device bandwidth of 18 GHz. We expect that by optimization of the waveguide geometry and of the doping profile in the slabs, the bandwidth of future SOH devices can be significantly increased [86], leading to considerably lower BER.

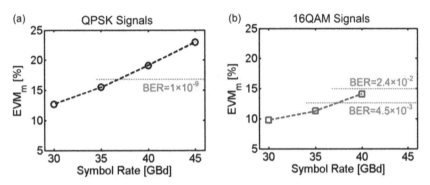

Fig. 3.16: EVM_m plotted over the symbol rate, in red the corresponding BER [77] is indicated. (a) Plot for QPSK signals. Up to 35 GBd the signal can be considered error free with a corresponding BER $< 10^{-9}$. The BER limit for hard-decision FEC corresponds to an EVM_m of 38.3 %, outside the range of the plot. (b) Plot for 16QAM signals. The horizontal dashed lines indicate the EVM_m that correspond to the BER thresholds for soft-decision and hard-decision FEC in the case of 16QAM.

When using a short SOH modulator at low symbol rates, the device can be operated without a terminal resistance and hence treated as a lumped capacitive load [17]. This is not possible for the high-speed data signals investigated here, for which the symbol duration is of the same order of magnitude as the propagation delay of the RF wave in the device. In this case, we need to model the modulator as a transmission line with a 50 Ω wave impedance, which is fed by a 50 Ω probe and terminated by a matched 50 Ω resistor, Fig. 3.17. This configuration is equivalent to a single 50 Ω load resistor, which is directly connected to the drive amplifier. The power consumption of a single MZM is then given by the power dissipation in this resistor. The output port of the drive amplifier is represented by an ideal voltage source with an internal generator impedance of 50 Ω. The energy per bit for the IQ modulator is obtained by adding the power consumptions of the two MZM and dividing by the total data rate. For 16QAM generation, the electrical drive signal for each MZM consists of two amplitude levels, which differ by a factor of 1/3. Assuming equal probability of the various symbols, the mean power consumption for each MZM can be calculated by averaging over the dissipated power of both amplitude levels. The energy per bit $W_{\text{bit,16QAM}}$ is then obtained by taking into account the power consumption of both MZM and by dividing by the data rate. Denoting the peak-to-peak voltages at the input of the $R = 50\,\Omega$ device as U_d, this leads to

$$W_{\text{bit,16QAM}} = 2 \times \frac{1}{2}\left[\left(\frac{U_d}{2}\right)^2 \frac{1}{R} + \left(\frac{1}{3}\frac{U_d}{2}\right)^2 \frac{1}{R}\right] \times \frac{1}{r_{\text{16QAM}}}. \qquad (3.35)$$

For $U_d = 1.80$ V and a data rate of $r_{\text{16QAM}} = 160$ Gbit/s, the energy consumption of the modulator is hence found to be 113 fJ/bit. This is one magnitude below current all-silicon 16QAM modulators [11], but still significantly higher than the energy consumption of a 28 GBd, 16QAM SOH device, for which 19 fJ/bit have been demonstrated [67]. It is the goal of ongoing research activity to further decrease the power consumption at high symbol rates.

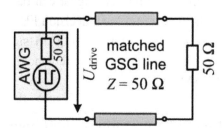

Fig. 3.17: Equivalent-circuit diagram of one MZM for calculation of the energy consumption. The AWG which drives the MZM is represented by an ideal voltage source and an internal resistance of 50 Ω. The GSG line of the MZM is matched to the 50 Ω output of the DAC and terminated by an external 50 Ω impedance. To estimate the energy consumption, the transmission line and its termination is replaced by an equivalent resistor of $R = 50\,\Omega$.

4. Summary

We experimentally demonstrate that SOH integration is capable of boosting data rates and energy efficiency of silicon-based IQ modulators to unprecedented values. We show 16QAM modulation at 40 GBd, resulting in line rates (net data rates) of 160 Gbit/s (133.3 Gbit/s), and QPSK modulation at 45 GBd, leading to 90 Gbit/s (84 Gbit/s). For 16QAM, the energy consumption is as low as 113 fJ/bit - an order of magnitude below that of comparable all-silicon devices.

[end of paper]

4 Frequency shifter integrated on the silicon platform

4.1 Frequency shifting in waveguide based modulators

The term frequency shifting refers to the ability to change the frequency of sinusoidal electro-magnetic waves, ideally without adding additional frequencies or changing the amplitude of the original wave too much. For optical signals often this is achieved via an acousto-optic modulator (AOM). This device uses a transparent material where a sound wave is induced, typically via a piezoelectric transducer, resulting in periodic index changes throughout the transparent material. An optical wave travelling through the transparent material now experiences Bragg reflection at the induced lattice and the Doppler shift changes the frequency of the optical wave [90]. AOMs provide a very clean frequency shift without spurious frequencies and are therefore widely used in different applications. However, this concept has two main drawbacks: Since the AOM is based on a mechanical wave propagating through a transparent medium, the speed is limited by the transducer and only frequency shifts of up to a few hundred MHz can typically be achieved. Furthermore, the AOM requires precise alignment in an optical free-space setup which leads to a large footprint and makes the whole setup susceptible to external disturbances. Optical integration is well suited to circumvent those problems. Photonic integrated circuits offer a multitude of functionalities on a very small footprint and can easily be used for high-speed operation. However, a direct translation of the operation principle of an AOM to common integrated optical circuits is not possible since on typical integration platforms piezo-electric materials are not available. Therefore, in the following chapter phase modulators, a basic building block of photonic integrated circuits, are used to achieve frequency shifting on the silicon photonic platform. To realize a frequency shifter with high signal quality and a large frequency shift, SOH phase modulators are used in a single-sideband modulation scheme with additional temporal shaping of the drive signal.

4.1.1 Single-sideband modulation for frequency shifting

Single-sideband (SSB) modulation is a transmission format used widely in radio communication in order to save spectral bandwidth and reduce power requirements on the transmitter [91]. It is characterized by a baseband signal which is modulated onto a carrier but compared to standard amplitude or frequency modulation, the transmitted signal carries only one sideband in the frequency domain, while the second sideband and usually also the carrier is suppressed. In the special case when the modulating signal is a sinusoidal, this relates to a frequency shift of the carrier by the frequency of the modulating signal either towards higher or lower frequencies.

SSB signals can be obtained by filtering the spectrum of an amplitude or frequency modulation signal. However, in the optical domain with carrier frequencies around 200 THz, implementation of such filters become impractical for modulation frequencies even in the GHz range. Especially for integrated optics the generation of SSB signals by phase modulators is preferred. In the following section the mathematical relations for SSB modulation and frequency shifting based on SSB modulation are described, the explanation is oriented at the derivations in [91] and [22].

A single-sideband signal can be represented using an analytical baseband signal

$$s_a = s_{BB} + j\hat{s}_{BB}, \quad (4.1)$$

where s_{BB} is a real valued baseband signal and $\hat{s}_{BB}$ the Hilbert transform of the baseband signal, see Appendix A.2. It is important to note that the Fourier transform of an analytical signal has only non-zero values for positive frequencies. The SSB signal can be obtained when the analytical signal is modulated onto a complex carrier with angular frequency ω_0,

$$s_{SSB} = s_a e^{j\omega_0 t} = s_{BB} e^{j\omega_0 t} + \hat{s}_{BB} e^{j(\omega_0 t + \pi/2)}. \quad (4.2)$$

The real valued modulation signal can be obtained by

$$\text{Re}\{\underline{s}_{\text{SSB}}\} = s_{\text{BB}}\cos(\omega_0 t) - \hat{s}_{\text{BB}}\sin(\omega_0 t). \quad (4.3)$$

If now a single sinusoidal signal $a_0 \sin(\Omega t)$ with modulation frequency Ω and amplitude a_0 is used as the amplitude modulated baseband signal, Eq. (4.3) translates to

$$\begin{aligned}\text{Re}\{\underline{s}_{\text{SSB}}\} &= a_0 \sin(\Omega t)\cos(\omega_0 t) + a_0 \cos(\Omega t)\sin(\omega_0 t) \\ &= a_0 \sin((\Omega + \omega_0)t).\end{aligned} \quad (4.4)$$

It can be seen in Eq. (4.4) that with SSB modulation the carrier frequency can be directly shifted by the modulation frequency Ω. It has to be noted that the sign of the frequency shift is arbitrary and depends only on the relative phase of the carrier to the sinusoidal modulation signal.

From Eq. (4.4) it is directly visible, that the frequency shift can be obtained by amplitude modulation of the in-phase and quadrature component of the carrier. This requires two sinusoidal signals with the same modulation frequency and $\pi/2$ phase difference. Such a modulator, also called Hartley modulator [92], is widely used in RF signaling. However, for integrated optics pure amplitude modulators are difficult to implement. Instead phase modulators are widely available and within a Mach-Zehnder configuration operated in push-pull mode, the phase modulation translates into an amplitude modulation.

4.1.2 Single-sideband modulation with phase modulators

To obtain a frequency shifted signal based on phase modulators, four phase modulators are combined in a nested IQ configuration as described in Section 2.2.2, see Fig. 4.1. The incoming complex signal with the carrier frequency ω_0 is split and fed with a $\pi/2$ phase difference into two Mach-Zehnder modulators operated in push pull configuration. The complex electrical field of the carrier at the input can be represented by

$$E_{in} = E_0 e^{j\omega_0 t}, \quad (4.5)$$

where E_0 is the field amplitude. The phase modulators are driven with a sinusoidal signal in push-pull mode, resulting in a phase modulation of

4 Frequency shifter integrated on the silicon platform

$$\Phi = a_0 \sin(\Omega t) \quad \text{and} \quad \Phi = -b_0 \cos(\Omega t), \tag{4.6}$$

where a_0 and b_0 are the modulation amplitude. The modulation frequency Ω is the angular frequency by which the optical carrier is shifted. After the modulator structure the field of the output signal is

$$E_{\text{out}} = \frac{1}{4} E_{\text{in}} \left(e^{ja_0 \sin(\Omega t)} - e^{-ja_0 \sin(\Omega t)} + je^{-jb_0 \cos(\Omega t)} - je^{jb_0 \cos(\Omega t)} \right). \tag{4.7}$$

For an ideal modulator we can assume $a_0 = b_0$, furthermore, by using the Jacobi-Anger expansion, Eq. (4.7) can be written as

$$E_{\text{out}} = \sum_{m=4n+1} E_0 J_m(a_0) e^{j(\omega_0 + m\Omega)t} \quad \text{for } n \in \mathbb{Z}. \tag{4.8}$$

Details on the calculation can be found in Appendix D.1. To maximize the field of the spectral line at $e^{j(\omega_0 + \Omega)t}$ the modulation amplitude a_0 of the phase modulator should be chosen such, that $J_1(a_0)$ is maximized, which is the case at $\max(J_1(a_0)) = J_1(a_0 = 1.841) = 0.582$. Assuming the modulation amplitude a_0 does not exceed this value, we can neglect higher order Bessel functions $J_m(a_0)$ for $|m| > 7$ since they deliver no significant contribution to the sum.

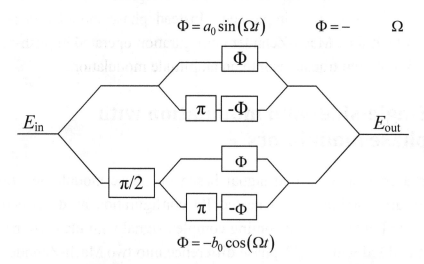

Fig. 4.1: Schematic of the modulator structure. To obtain a frequency shift with the SOH phase modulators, two each pair of the phase shifter is used as a Mach-Zehnder modulator (MZM) in push-pull operation. In the optical path there is a relative phase difference of π/2 between both MZM.

4.1.3 Small signal modulation and temporal shaping for enhanced side-mode suppression ratio

As derived in the previous section, the output signal of a frequency shifter based on nested phase modulators consists of several spectral components at frequencies $(\omega_0 + m\Omega)$. Besides the desired frequency component at $(\omega_0 \pm \Omega)$ the other modes are spurious side modes which are normally unwanted. As derived in Eq. (4.4), perfect frequency shifting without spurious side modes can be obtained by an sinusoidal amplitude modulation of the in-phase and quadrature component of the carrier. The structure in Fig. 4.1 relies on Mach-Zehnder modulators in push-pull operation for amplitude modulation. Those MZM feature a sinusoidal transfer function, see Section 2.2.1. When operating the phase shifter with a sinusoidal modulation, the resulting amplitude modulation becomes a distorted sinusoidal. This give rise to the spurious side modes in Eq. (4.8).

Depending on the modulation amplitude a_0 the contribution of the different Bessel functions to the output signal are varying. To obtain a metric on the quality of the output signal it is useful to calculate the ratio between the frequency components of the output signal in dependence of the modulation amplitude. Using Eq. (4.8) we can write the ratio in terms of the Bessel functions

$$\left| \frac{E_{\text{out},\omega_0+m\Omega}(a_0)}{E_{\text{out},\omega_0+\Omega}(a_0)} \right| = \frac{|J_m(a_0)|}{|J_1(a_0)|} \quad \text{for } m = 4n+1,\ n \in \mathbb{Z}\setminus 0. \quad (4.9)$$

A plot of the Bessel functions J_1, J_3, J_5 and J_7, as well as the ratios J_3/J_1, J_5/J_1 and J_7/J_1 can be seen in Fig. 4.2. For larger modulation amplitudes the contributions of the higher order Bessel functions become larger. The dominant spurious side mode is at $(\omega_0 - 3\Omega)$ since the third order Bessel function grows faster than the higher order Bessel functions. For $a_0 = 1.841$ the frequency conversion into the mode at $(\omega_0 + \Omega)$ is largest, however, the ratio with the spurious side mode is with $J_3/J_1 = 0.18$ relatively large. Since normally spectral intensity is measured we can use the ratio of the intensities instead of the field ratio as figure of merit. We define the side mode suppression ratio (SMSR) as the ratio between the intensity of the signal shifted by Ω and the intensity of the largest spurious side mode.

4 Frequency shifter integrated on the silicon platform

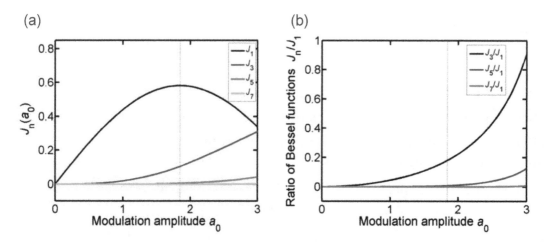

Fig. 4.2: (a) Plot of the Bessel functions over the modulation amplitude up to the 7th order. The modulation amplitude, where the maximum of the first order Bessel function is reached is marked with a dotted line. (b) Ratio of the different Bessel functions with respect to the first order Bessel function.

$$\text{SMSR} = \frac{I_{\text{out},\omega_0+\Omega}}{\max\left(I_{\text{out},\omega_0+m\Omega}\right)}$$
$$= \frac{\left|E_{\text{out},\omega_0+\Omega}\right|^2}{\max\left(\left|E_{\text{out},\omega_0+m\Omega}\right|^2\right)} = \frac{\left|J_1(a_0)\right|^2}{\left|J_3(a_0)\right|^2} \quad \text{for } m = 4n+1, n \in \mathbb{Z} \setminus 0 \quad (4.10)$$

For smaller modulation amplitudes the SMSR becomes large due to the fact, that only the linear part of the MZM transfer function is used and therefore the output of the frequency shifter can be approximated as sinusoidal amplitude modulation.

Alternatively, the nonlinear MZM transfer function can be compensated by proper temporal shaping, or predistortion, of the modulation signal Φ and $\bar{\Phi}$. This can be achieved by adding additional harmonic frequencies to the modulation. Exemplarily the compensation of the spectral component at $(\omega_0 + 3\Omega)$ will be shown here. By adding the third harmonic to the modulation signal we obtain

$$\Phi = a_0 \sin(\Omega t) - c_0 \sin(3\Omega t) \quad \text{and} \quad \bar{\Phi} = -a_0 \cos(\Omega t) - c_0 \cos(3\Omega t) \quad (4.11)$$

4.1 Frequency shifting in waveguide based modulators

and the field at the output changes to

$$E_{out} = \frac{1}{4}E_{in} \begin{pmatrix} e^{ja_0 \sin(\Omega t)}e^{-jc_0 \sin(3\Omega t)} - e^{-ja_0 \sin(\Omega t)}e^{jc_0 \sin(3\Omega t)} \\ +je^{-ja_0 \cos(\Omega t)}e^{-jc_0 \cos(3\Omega t)} - je^{ja_0 \cos(\Omega t)}e^{jc_0 \cos(3\Omega t)} \end{pmatrix}. \quad (4.12)$$

As described in Appendix D.2, this equation can be simplified to

$$E_{out} = E_0 \begin{pmatrix} J_0(c_0) \sum_{m=4n+1} J_m(a_0)e^{j(\omega_0+m\Omega)t} \\ -J_0(a_0) \sum_{l=4k-1} J_l(c_0)e^{j(\omega_0+l3\Omega)t} \end{pmatrix}, \quad (4.13)$$

with the significant components of the sum described by

$$E_{out} = \begin{pmatrix} \ldots + J_0(c_0)J_1(a_0)e^{j(\omega_0+\Omega)t} + J_0(c_0)J_{-3}(a_0)e^{j(\omega_0-3\Omega)t} \\ -J_0(a_0)J_{-1}(c_0)e^{j(\omega_0-3\Omega)t} + J_0(c_0)J_5(a_0)e^{j(\omega_0+5\Omega)t} + \ldots \end{pmatrix}. \quad (4.14)$$

With a proper choice of a_0 and c_0 the majority of the field at the frequency $(\omega_0 - 3\Omega)$ can be suppressed by fulfilling the equation

$$\begin{aligned} J_0(c_0)J_{-3}(a_0)e^{j(\omega_0-3\Omega)t} - J_0(a_0)J_{-1}(c_0)e^{j(\omega_0-3\Omega)t} &= 0 \\ J_{-3}(a_0) &= J_{-1}(c_0) \end{aligned}. \quad (4.15)$$

This concept can be expanded for the higher harmonics present in the output spectrum. Using an infinite number of additional harmonics in the modulation signal, in principle all side modes can be suppressed and only the shifted component at $(\omega_0 + \Omega)$ is retained.

For a more intuitive explanation we can recall the sinusoidal transfer function of a Mach-Zehnder modulator, see Section 2.2.1. In order to obtain ideal frequency shifting, the resulting amplitude modulation of each MZM must be a sinusoidal, therefore a proper drive signal has to be applied. In the particular case, when the modulator is driven with a triangular signal with an amplitude of $a_0 = \pi/2$, the output of the MZM is exactly a sinusoidal amplitude modulation. The output of the frequency shifter is then an ideally shifted line with the field

$$E_{out} = \frac{1}{2}E_0 e^{j(\omega_0+\Omega)t} \quad (4.16)$$

for the given structure in Fig. 4.1. The factor 0.5 in Eq. (4.16) comes from the fact that in each coupler half the power is routed in the destructive port when combining the modulated signals.

4.1.4 Frequency shifting via serrodyne modulation

Serrodyne frequency shifting is based on the idea that for a wave E with frequency ω and amplitude E_0 a continuous linear phase shift $\phi(t) = \Omega \cdot t$ corresponds to a frequency shift according to

$$E = E_0 e^{j(\omega t + \phi(t))} = E_0 e^{j(\omega + \Omega)t}. \qquad (4.17)$$

To realize such a quasi-continuous phase shift in practical applications a sawtooth phase shift is used. The phase is ramped perfectly linear from 0 to $2m\pi$ with $m \in \mathbb{Z}$ and has an ideally infinite short fall time T_f [21,93]. If the period of the sawtooth signal is T, the linear slope can be described by $\phi_0 = 2m\pi/T$. Following the derivation in [21] the output of a serrodyne modulator can be described as

$$E = E_0 \left[e^{j\omega t} \cdot \sum a_m e^{j\frac{2\pi m}{T}t} \right], \qquad (4.18)$$

where a_m is the amplitude of the mode which is shifted by m/T. However, in reality an ideal sawtooth signal with infinite short fall time cannot be realized which leads to distortions in the output signal. An illustration of an ideal serrodyne signal $\phi_{\text{ideal}}(t)$ and a real signal $\phi_{\text{real}}(t)$ is depicted in Fig. 4.3. As described in [21], the amplitude a_m of the side modes can be evaluated via

$$|a_m| = \left| \frac{1}{T} \int_0^T \cos\left(\phi_{\text{real}}(t) - \frac{2\pi m}{T} \right) dt \right|. \qquad (4.19)$$

This leads to the finding that for achieving a side mode suppression ratio larger 30 dB, the fall time must be less than 3% of the period of the serrodyne signal. This sets substantial requirements for the signal source, generating the serrodyne signal. A serrodyne signal can be described with the Fourier series

$$\phi(t) = \frac{2}{\pi} \sum_{n=1}^{\infty} \frac{(-1)^{n-1}}{n} \sin\left(\frac{\pi n t}{T} \right), \qquad (4.20)$$

where T is again the period of the serrodyne signal. To achieve a fall time of $T_f/T \leq 3\%$ harmonic frequencies with frequencies more than 30 times higher than the fundamental frequency of the serrodyne signal must be present in the signal.

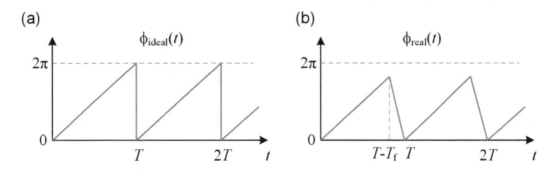

Fig. 4.3: (a) Illustration of an ideal sawtooth signal with period T and. The linear slope reaches from 0 to 2π. (b) Approximation of a real sawtooth signal with fall time T_f, the slope of the signal is the same as for the ideal signal.

4.2 State of the art

To realize frequency shifters in photonic integrated circuits typically phase shifters are used as a basic building block. As discussed in Section 4.1 both, single-sideband modulation and serrodyne modulation can be employed to achieve the desired frequency shift.

The LiNbO$_3$ platform is well suited for integrated frequency shifter based on phase modulation since its inherent $\chi^{(2)}$-nonlinearity allows for pure phase modulation. Based on the single-sideband modulation, several devices were demonstrated with phase shifts at high frequencies. Shimotsu et al., demonstrated a frequency shift of 10 GHz with a SMSR of 18.4 dB [94]. Also serrodyne frequency shifter have been demonstrated integrated on LiNbO$_3$ [95]. Poberezhskiy et al. demonstrated a frequency shift of 1.28 GHz using a LiNbO$_3$ phase modulator. To achieve such a large frequency shift with serrodyne modulation, the broadband serrodyne drive signal is generated via a complex RF-photonic arbitrary waveform generator. However, the resulting SMSR is with 12 dB relatively low. Both approaches share the problem of high drive voltages when relying on the LiNbO$_3$ platform. The $\chi^{(2)}$ nonlinearity within the crystal is relatively weak and leads therefore to high drive voltages in the order of 5 V – 10 V. Furthermore, the footprint of a LiNbO$_3$ PIC is relatively large due to its weak index contrast. Modulator lengths are typically in the order of 1 cm.

The silicon photonic platform with its high index contrast allows dense integration. There are several ways to realize phase modulators with silicon photonics. For optical communication phase modulators based on the plasma dispersion effect are often used. However, they do not allow for pure phase modulation since a change in the carrier density comes with an associated amplitude modulation which is detrimental for a high signal quality in frequency shifters. The SMSR of both, serrodyne modulation and single-sideband modulation suffer from any residual amplitude modulation [96]. Up to now, frequency shifter integrated on SOI are therefore realized with thermal phase shifters. Those thermal phase shifters exploit the strong thermo-optic coefficient in silicon to change the refractive index of a waveguide and thereby the phase of a propagating wave in the waveguide. Such phase shifters are very compact, below 100 um in length [97]. However, thermal phase shifters have an inherently low bandwidth due to the fact that heat transfer in the device is comparatively slow. Typical bandwidths of such devices are in the order of few hundred kHz [97,98]. The frequency shifters realized so far on silicon with thermal phase shifters are using the serrodyne modulation. They show a conversion efficiency close to unity and a SMSR as high as 39 dB [23,24]. However, due to the slow thermo-optic phase modulators, the achievable frequency shift is limited to few ten kHz.

A goal of this work is to demonstrate a frequency shifter which leverages all the benefits of the silicon photonic platform and allows for large frequency shifts. Furthermore, relaxed requirements on the electrical drive signals are targeted. In the following section a frequency shifter is demonstrated using silicon-organic hybrid integration. The pure phase modulation and the high efficiency of the SOH phase modulators makes them especially suitable for the application in integrated frequency shifters.

4.3 Silicon-organic hybrid frequency shifter

Using a silicon-organic hybrid IQ modulator, the concept of a single-sideband frequency shifter together with temporal shaping of the drive signal was demonstrated on the silicon platform. This section was published in a scientific journal [J17].

[start of paper]

Integrated optical frequency shifter in silicon-organic hybrid (SOH) technology

Optics Express Vol. 24, Issue 11, pp. 11694-11707 (2016)

DOI: 10.1364/OE.24.011694 © 2016 Optical Society of America

M. Lauermann,[1] C. Weimann,[1] A. Knopf,[2] W. Heni,[1,6] R. Palmer,[1,7] S. Koeber,[1,5,8] D. L. Elder,[3] W. Bogaerts,[4] J. Leuthold,[1,6] L. R. Dalton,[3] C. Rembe,[2,9] W. Freude,[1] C. Koos[1,5]

[1] *Karlsruhe Institute of Technology, Institute of Photonics and Quantum Electronics, 76131 Karlsruhe, Germany*

[2] *Polytec GmbH, 76337 Waldbronn, Germany*

[3] *University of Washington, Department of Chemistry, Seattle, WA 98195-1700, United States*

[4] *Ghent University – IMEC, Photonics Research Group, Gent, Belgium*

[5] *Karlsruhe Institute of Technology, Institute of Microstructure Technology, 76344 Eggenstein-Leopoldshafen, Germany*

[6] *Now with: Institute of Electromagnetic Fields, Swiss Federal Institute of Technology (ETH), Zurich, Switzerland*

[7] *Now with: Coriant GmbH, 81541 Munich, Germany*

[8] *Now with University of Cologne, Chemistry Department, 50939 Cologne, Germany*

[9] *Now with: Clausthal University of Technology, Institute of Electrical Information Technology, 38678 Clausthal-Zellerfeld, Germany.*

We demonstrate for the first time a waveguide-based frequency shifter on the silicon photonic platform using single-sideband modulation. The device is based on silicon-organic hybrid (SOH) electro-optic modulators, which combine conventional silicon-on-insulator waveguides with highly efficient electro-optic cladding materials. Using small-signal modulation, we demonstrate frequency shifts of up to 10 GHz. We further show large-signal modulation with optimized waveforms, enabling a conversion efficiency of -5.8 dB while suppressing spurious side-modes by more than 23 dB. In contrast to conventional acousto-optic frequency shifters, our devices lend themselves to large-scale integration on silicon substrates, while enabling frequency shifts that are several orders of magnitude larger than those demonstrated with all-silicon serrodyne devices.

1. Introduction

Frequency shifters are key elements for a wide range of applications, comprising, e.g., heterodyne interferometry, vibrometry, distance metrology, or optical data transmission [99–102]. All of these applications can benefit greatly from photonic integration, which allows realization of optical systems with high technical complexity while maintaining robustness, small footprint, and low cost. Silicon photonics is a particularly attractive platform for optical integration, leveraging large-scale, high-yield CMOS processing and offering the potential of co-integrating photonic and electronic devices on a common substrate [78]. Driven by applications in optical communications, a variety of foundry services have emerged over the last years offering process design kits (PDK) and libraries of standardized silicon photonic devices such as passive components, electro-optic modulators, and Germanium-based photodetectors [8]. High-performance frequency shifters, however, are still missing in the portfolio, especially when it comes to devices that combine high conversion efficiency (CE) between the original and the frequency-shifted signal with a high side-mode suppression ratio (SMSR) of the original carrier and the spurious side modes relative to the shifted spectral line.

Conventionally, frequency shifters have been realized with acousto-optic modulators (AOM), exploiting the Doppler shift of an optical wave that is scattered from a propagating sound wave [90]. When aligned for perfect phase matching, AOM provide both high SMSR and CE. However, the alignment is delicate and needs to be distinctively linked to the desired frequency shift. In addition, AOM are based on free-space optical assemblies of conventional high-precision optomechanical components [103], making the devices bulky, expensive, and sensitive to vibrations and environmental influences. Moreover, AOM require material systems with strong acousto-optic and ideally also piezoelectric effects, ruling out silicon as an integration platform by fundamental principles. AOM-based systems are hence inherently unsuited for integration on the silicon platform.

Alternatively, waveguide-based frequency shifters can be used, exploiting either the so-called serrodyne technique [21] or single-sideband modulation in an in-phase/quadrature (IQ) modulator [22]. For serrodyne modulation, an electro-optic phase shifter is driven by a sawtooth signal with a peak-to-peak drive amplitude that leads exactly to a 2π phase shift of the optical carrier. This results in a phase shift that is piecewise linear in time and hence leads to a frequency shift that corresponds to the fundamental frequency of the sawtooth drive signal. Such devices have been demonstrated on the silicon photonic platform, providing a remarkable SMSR of 39 dB [23,24], but the achievable frequency shift is limited by fundamental properties of silicon-based phase shifters: To avoid unwanted spurious lines in the output spectrum, pure serrodyne phase modulation must be achieved without any residual amplitude modulation. This cannot be accomplished by using conventional silicon photonic phase shifters that rely on injection or depletion of free carriers in reverse or forward-biased p-n or p-i-n junctions [7]. Instead, thermo-optic phase modulators had to be used, leading to stringent limitations of the achievable frequency shift. In particular, a good SMSR requires a perfect sawtooth-like phase shift that contains a large number of higher-order harmonics besides the fundamental frequency component [21]. Given the time constants and the associated bandwidth limitations of thermo-optic phase shifters [104], this is possible for very small fundamental frequencies only, thereby limiting the achievable frequency shifts to values of, e.g., 10 kHz [23], which is insufficient for many applications. In the case of single-sideband (SSB) modulation, two Mach-Zehnder modulators (MZM) are combined with a $\pi/2$ phase shift to form an optical IQ modulator [22]. Both MZM are biased in the zero-transmission point. The in-phase (I) and quadrature (Q) MZM are driven with a sine signal and cosine signal, respectively, which, in the small-signal regime, leads to a frequency shift that corresponds to the frequency of the sinusoidal drive signals. As for the serrodyne method, a good SMSR requires a pure frequency shift without any spurious amplitude-phase coupling, making it challenging to realize high-performance single-sideband modulators based on the standard silicon photonic device portfolio. Previous demonstrations of integrated optical SSB frequency shifters are therefore limited to the LiNbO$_3$ platform [94] which is not suited for large-scale photonic integration and needs high drive voltages of more than 6 V$_{pp}$. Moreover, conventional SSB modulation suffers from a trade-off between SMSR and CE: For high SMSR, the MZM need to be driven in the small-signal regime, where the amplitude transmission is proportional to the applied voltage. This leads to large modulation loss and hence small CE. High CE, on the other hand, would require large signals to drive the device to maximum transmission, thereby inducing spurious side-modes in the optical spectrum due to the nonlinear transfer function of the MZM [94].

In this paper we demonstrate the first single-sideband (SSB) frequency shifter based on silicon photonic waveguides. We exploit the silicon-organic hybrid (SOH) integration concept to realize broadband phase shifters that feature low drive voltages and enable pure phase modulation. SOH integration permits the combination of conventional silicon-on-insulator (SOI) waveguides with a wide variety of organic cladding materials and has been used to realize efficient high-speed electro-optic modulators [56,17,14], ultra-compact phase shifters [105,106], as well as lasers [107]. Interaction of the guided light with the organic cladding of the SOH device can be enhanced by using thin ultra-thin strip [108] or slot waveguides [82], which can be combined with photonic crystal structures [64,109] or ring resonators [110]. In our experiments, we use conventional SOH slot-waveguide modulators to show frequency shifts of up to 10 GHz and a carrier suppression of 37 dB for the case of SSB modulation in the small-signal regime [111]. Moreover, we experimentally demonstrate temporal shaping of the drive signal to overcome the trade-off of SMSR and CE. We demonstrate a CE of -5.8 dB while maintaining a SMSR of 23.5 dB. The experimentally obtained CE is quite close to the theoretically predicted value of -6 dB. These results are achieved with first-generation devices, leaving considerable room for further improvement.

This paper is structured as follows: In Section 2, we explain the theoretical concepts of frequency shifters based on IQ modulators. Section 3 introduces the SOH phase shifters used for realizing the frequency shifters. Section 4 describes experimental demonstrations of the devices and discusses the achieved performance. A rigorous mathematical description of frequency shifters based on single-sideband modulation is given in the Appendix.

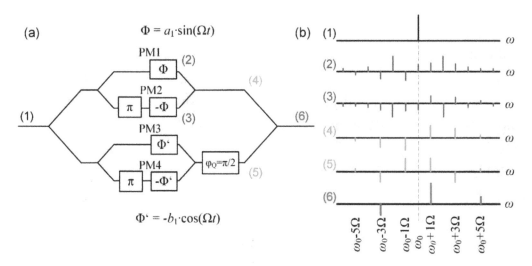

Fig. 4.4: (a) IQ modulator structure for single-sideband operation. **(a)** Schematic of the modulator structure. Four phase modulators (PM 1 … 4) are paired to two Mach-Zehnder modulators (MZM). The MZM are biased in zero transmission and operated in push-pull mode with sinusoidal electrical drive signals that feature a relative phase shift of $\pi/2$. The outputs of the MZM are combined with a relative optical phase difference of $\pi/2$. **(b)** Sketch of the field amplitude spectrum at various points of the modulator structure. The spectral lines feature phases of 0 or π, which are indicated by positive or negative peak values. At the output of the frequency shifter, most of the energy has been transferred into the line that is shifted by the modulation frequency Ω. When the MZM is not driven in the small-signal regime, additional harmonics can be seen in the output spectrum. Ω represents the RF modulation frequency while ω_0 denotes the frequency of the optical carrier.

2. Frequency shifters based on IQ-modulators

The concept of SSB frequency shifting with an integrated IQ modulator was first described by Izutsu *et al.* in 1981 [22]. The general principle is explained in Fig. 4.4 (a): The optical input is split and fed into four phase modulators which are paired to two MZM, the outputs of which are combined with a relative phase shift of $\pi/2$. Both MZM are biased at zero transmission (null point), indicated by the π phase shift in the lower arm of each MZM in Fig. 4.4(a). The devices are driven in push-pull mode by a sine and cosine of angular frequency Ω, respectively. Fig. 4.4(b) illustrates the optical spectra that can be found at various positions (1) … (6) in the SSB modulator setup. After one phase shifter, a multitude of spectral lines emerge in the frequency domain, separated by the modulation frequency Ω, see displays (2) and (3) in Fig. 4.4(b). The amplitudes of the lines depend on the modulation depth and can be described by Bessel functions, see Appendix for a detailed mathematical description. When the signals of a pair of phase shifters are combined at the output of the MZM, the spectral components at the optical carrier frequency ω_0 and all even-order side lines interfere destructively, see displays (4) and (5). Moreover, at the output of the IQ modulator, the signals from both MZM are combined with a relative phase shift of $\pi/2$, leading to a destructive interference of the lines at $\omega_0 - 1\Omega$, $\omega_0 + 3\Omega$, $\omega_0 - 5\Omega$ etc., see display (6) in Fig. 4.4(b). The direction of the frequency shift can be chosen arbitrarily: By assigning the $\pi/2$ phase shift to the other arm of the IQ modulator, the lines interfere constructively at $\omega_0 - 1\Omega$ instead of $\omega_0 + 1\Omega$, the output spectrum is flipped about ω_0. For practically relevant modulation depths, the spectral line at $\omega_0 + \Omega$ carries most of the energy at the output of the frequency shifter, while unwanted side modes appear at $\omega_0 - 3\Omega$, $\omega + 5\Omega$, $\omega - 7\Omega$,... with decreasing intensity, see Appendix for a mathematical description.

For small-signal modulation, i.e., for drive voltage amplitudes that are much smaller than the π-voltage of the MZM, the spurious side modes at $\omega_0 - 3\Omega$, $\omega_0 + 5\Omega$ and higher can be neglected. This can be understood intuitively when considering the field amplitude transfer function of a single MZM, which is depicted in Fig. 4.5 along with the corresponding output spectrum for different drive conditions. For small-signal modulation around the null point, only the linear part of the MZM field transfer function is used for modulation by the sinusoidal drive signal, hence the amplitude modulation at the output is sinusoidal as well. This results in only two side-modes, see Fig. 4.5(a). Since the MZM are biased in the null-point, small-signal modulation leads to a low conversion efficiency, and most of the power is lost to the destructive port of the amplitude combiners at the output of the MZM. For large-signal modulation, in contrast, the output signal is a distorted sine, leading to larger conversion efficiency but spurious side modes, Fig. 4.5(b).

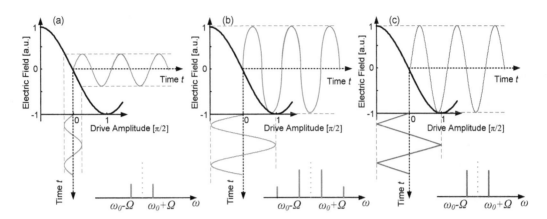

Fig. 4.5: Field amplitude transfer characteristics of an ideal MZM in push-pull operation along with the corresponding output spectra, the drive voltage and the electrical field at the output for different drive conditions. (a) Small-signal operation: The MZM is driven with a sinusoidal signal in the linear regime, resulting in two weak lines at $\omega_0 \pm \Omega$ in the output spectrum. The amplitude of the output field is relatively small, leading to a low conversion efficiency. The carrier frequency (dotted line) is only depicted for reference – it should not contain any power after the MZM, which is biased in the null point for zero transmission. (b) Large-signal operation with a sinusoidal drive signal. The field at the output resembles a distorted sine with large amplitude. Therefore the desired spectral lines have large intensity, but are accompanied by spurious side modes. (c) Large-signal operation with a triangular drive signal. For an ideal MZM, the output signal features a perfectly sinusoidal amplitude modulation, and the spectrum consists of only two lines at $\omega_0 \pm \Omega$. – this time with large intensity.

To achieve high conversion efficiency and high SMSR simultaneously, the drive signal has to be predistorted to suppress the spurious side modes. This can be accomplished by adding higher harmonics to the drive signal to generate new side modes, which can destructively interfere with the spurious modes from the fundamental drive frequency. For example, when the third harmonic $a_3 \sin(3\Omega)$ is added to the RF drive signal, the optical amplitude at the output can be described by

$$E = E_0 e^{j\omega_0 t} \begin{pmatrix} \ldots + J_0(a_3)J_1(a_1)e^{j\Omega t} \\ + J_0(a_3)J_{-3}(a_1)e^{-j3\Omega t} - J_0(a_1)J_{-1}(a_3)e^{-j3\Omega t} \\ + J_0(a_3)J_5(a_1)e^{j5\Omega t} \\ + J_0(a_3)J_{-7}(a_1)e^{-j7\Omega t} + \ldots \end{pmatrix}, \quad (4.21)$$

where E is a complex amplitude factor, $J_n(a_1)$ describes the Bessel function of n-th order and a_1 (a_3) is the amplitude of the first (third) harmonic of the electrical drive signal. A derivation of this relation is explained in the Appendix. With the correct amplitude levels a_1 and a_3 for the harmonics of the drive signals, the relation $J_0(a_3)J_{-3}(a_1)e^{-j3\Omega t} = J_0(a_1)J_{-1}(a_3)e^{-j3\Omega t}$ can be fulfilled and the frequency component at $\omega_0 - 3\Omega$ can be removed from the output signal, except for the contributions including the higher-order Bessel functions $J_k(a_3)$, $|k| > 1$, which can be neglected for practical modulation depths. The same procedure can be applied to remove all other spurious side modes. Hence, using an infinite number of higher harmonics in the drive signal, we can in principle suppress all side modes and only retain the frequency-shifted component at $\omega_0 + \Omega$. For an ideal IQ modulator, the resulting drive signal would be perfectly triangular and feature an amplitude of $a_1 = \pi/2$.

An intuitive illustration of this situation is again depicted in Fig. 4.5(c): Given the sinusoidal transfer function of the MZM, a perfectly triangular drive signal driven with amplitude $a_1 = \pi/2$ results in a sinusoidal amplitude modulation at the output, leading to only two discrete side modes at the output, one of which is annihilated by destructive interference at the output coupler of the IQ modulator. This allows realizing an ideal frequency shifter without any spurious side modes. Compared to serrodyne modulation, the sinusoidal drive signals for SSB modulation are much easier to generate than the sawtooth signal, and even the triangular drive signal has significantly less power in higher harmonic frequencies than the sawtooth signal, thereby facilitating high-bandwidth operation of the frequency shifter. Note that the preceding considerations are valid for an ideal IQ modulator, without any amplitude imbalance or phase errors. For real devices, driving the SSB modulator with a triangular signal will not necessarily lead to a perfect frequency shift. Still, the harmonics of the drive signal can be adjusted to compensate also the imperfections of the modulator.

3. Silicon-organic hybrid (SOH) frequency shifters

To avoid amplitude-phase coupling in the silicon-based modulators, the phase shifter sections are realized by silicon-organic hybrid (SOH) integration [56]. A cross section of an SOH MZM is depicted in Fig. 4.6(a). Each phase shifter consists of a 1.5 mm long silicon slot waveguide (w_{slot} = 140 nm, w_{rail} = 220 nm, h_{Rail} = 220 nm) with an electro-optic organic material as cladding. The device is operated in quasi-TE polarization at a wavelength of 1540 nm. In this configuration, the high index contrast between the silicon rails (n = 3.48) and the organic cladding (n = 1.83) leads to an enhanced optical field in the slot. A thin n-doped silicon slab electrically connects the rails to ground-signal-ground (GSG) travelling-wave electrodes. The modulation RF voltage applied to the electrodes drops predominantly over the narrow slot, leading to a strong modulating RF field, which overlaps with the optical quasi-TE mode. This leads to highly efficient phase modulation. As a figure of merit for the overlap, the field confinement factor Γ quantifies the fraction of the optical field which interacts with the modulation field [18,38]. For our device the field confinement factor amounts to Γ = 0.24. The modulation efficiency of SOH devices can be increased by using a narrower slot, which leads to an increased field confinement factor and at the same time leads to a higher modulation field for a given drive voltage. We chose here a slot width of w_{slot} = 140 nm as a compromise between decent modulation efficiency and reliable fabrication using deep UV (DUV) lithography. A detailed discussion on the influence of the waveguide dimensions on the field confinement factor can be found in [56].

A gate voltage applied between the SOI substrate and the device layer can be used to induce an electron accumulation layer at the oxide-silicon interface, thereby increasing the conductivity of the silicon slabs [68,112]. In our experiment, the gate field is applied across the 2 µm thick buried oxide (BOX), which leads to a rather high voltage of 300 V to achieve a field strength of 0.15 V/nm. Using gate contacts on top of the silicon slab, the gate voltage can be reduced to less than 3 V [15]. For optimized doping profiles, the gate voltage can even be omitted completely – we have recently demonstrated high-speed operation of an SOH MZM operating at 80 Gbit/s without any applied gate voltage [113].

4 Frequency shifter integrated on the silicon platform

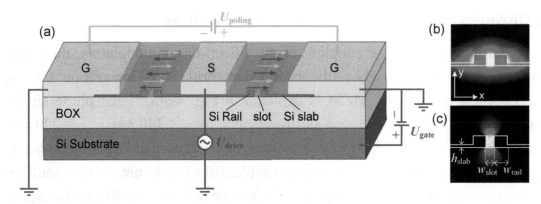

Fig. 4.6: (a) Cross-sectional view of an SOH MZM. The slot waveguides are filled with the electro-optic organic material DLD164. Electrical contact to the ground-signal-ground (GSG) transmission line of the MZM is established by thin *n*-doped silicon slabs. The EO material in the slot is poled via a DC poling voltage U_{poling} applied to the floating ground electrodes of the MZM, and the electro-optic chromophores in both arms orient along the electrical field in the same direction as indicated by green arrows. For operation of the device, an RF signal is fed to the GSG electrodes. The associated electric field, indicated by blue arrows, is parallel to the poling direction in the right arm, and antiparallel in the left arm. This leads to phase shifts of equal magnitude but opposite sign and hence to push-pull operation of the MZM. A static gate voltage U_{gate} is applied between the substrate and the device layer to increase the conductivity of the silicon slab. (b) Dominant E_x component of the optical field and (c) E_x component of the electrical drive signal in the slot waveguide. Both fields are confined to the slot and overlap strongly within the EO material, thereby enabling efficient phase shifting.

The optical cladding material consists of the strongly electro-optic chromophore DLD164 [17,114], which allows for pure phase modulation of the optical signal. The material is deposited by spin coating from 1,1,2-trichloroethane onto pre-fabricated SOI waveguide structures and dried in vacuum. To align the organic molecules, a poling process is used during fabrication: The device is heated to the glass transition temperature of 66 °C of the organic material, and a poling voltage is applied via the floating ground electrodes of each MZM. This results in a poling field of approximately 280 V/μm in each slot waveguide. The dipolar chromophore molecules hence align in the same direction as defined by the electric field, indicated by the green arrows in Fig. 4.6(a). After alignment, the sample is cooled down to preserve the alignment of the chromophores, and the poling voltage is removed. Using DLD164, a voltage-length product of $U_\pi L = 0.5$ Vmm is achieved, corresponding to an in-device electro-optic coefficient of $r_{33} = 150$ pm/V and a poling efficiency of 0.54 cm²/V². The in-device

r_{33} of the SOH devices is higher than the electro-optic coefficient of $r_{33} = 137$ pm/V achieved for thin film experiments with DLD164, for which poling voltages were limited by dielectric breakdown [114]. Those findings are in line with previous investigations where a high in-device electro-optic coefficient of up to $r_{33} = 180$ pm/V was achieved [17]. For operating the device, an RF signal is coupled to the GSG transmission line, leading to a modulating electric RF field as indicated by blue arrows in Fig. 4.6(a). The RF field and the chromophore orientation are antiparallel to each other in the left half of the GSG transmission line, and parallel in the right half. This leads to phase shifts of opposite sign in the two arms of the MZM and hence to push-pull modulation.

In the experiment, light is coupled to the chip via grating couplers and routed to the frequency shifter via standard SOI strip waveguides. Multimode interference couplers (MMI) are used as power splitters and combiners. To couple the light into the slot waveguide, optimized logarithmic tapers are used [71]. The fiber-to-fiber loss of the device amounts to 29 dB when tuning the modulator to maximum transmission. This rather high excess loss is dominated by the coupling loss of the two grating coupler with approx. 6 dB, each. The 7 mm long access waveguides add 0.6 dB/mm loss due to sidewall roughness. The optimized strip-to-slot converters and MMI have an aggregate loss of less than 2 dB. The 1.5 mm long active slot waveguide is estimated to have a loss of 7.3 dB/mm. Using optimized device design and fabrication processes, the overall insertion loss of the frequency shifter can be reduced considerably in future device generations: Access waveguides can be realized with losses below 1 dB, and optimized grating couplers and photonic wire bonds are available with fiber-chip coupling losses of less than 2 dB [76,115]. Using asymmetric slot waveguides [73] with improved doping profiles, the propagation loss in the modulator section can be reduced to below 1.5 dB/mm. This leads to an overall on-chip loss of less than 5 dB and to an overall fiber-to-fiber of less than 9 dB.

4. Experimental Demonstration

The setup used for experimental demonstration of the phase shifter is sketched in Fig. 4.7. A tunable laser is coupled to the silicon chip via grating couplers. On the chip, two SOH MZM are nested to form an IQ configuration. An intentional length imbalance of 40 µm is introduced between the two arms that contain the MZM such that the 90° phase shift between I and Q component can be adjusted by wavelength tuning. The electrical drive signal is fed to the chip via microwave probes, and the device is biased at the null point via bias-Ts. For our experiments, we usually chose the null point that features the lowest bias voltage. We also tried other null points without observing any relevant difference in device performance.

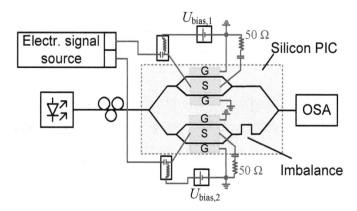

Fig. 4.7: Sketch of the measurement setup. A tunable laser is used as an optical source and coupled to the silicon photonic integrated circuit (PIC) via a grating coupler (GC). The PIC consists of two SOH MZM that are nested in an IQ configuration. An intentional imbalance of 40 µm is inserted between the two MZM to enable phase adjustment via wavelength tuning. For small-signal modulation with a purely sinusoidal electrical drive signal, we use a radio-frequency (RF) synthesizer featuring a bandwidth of 40 GHz as an electrical signal source. The RF-signal is split in two paths with a length difference of 20 cm such that the phase shift between the two drive signals can be adjusted by fine-tuning of the RF frequency. For large-signal operation with pre-distorted waveforms, an arbitrary-waveform generation (AWG) with two signal channels is used, where the waveform, output power, and phase can be individually adjusted for each channel. The bias voltages for the MZM are coupled to the chip via bias-Ts.

The ends of the transmission lines are terminated with 50 Ohm to avoid spurious reflections of the electrical signal. The RF performance of our device is limited by the 50 nm thin resistive silicon slabs, the RC limitations of which can be only partly overcome by the gate voltage. There is also an impedance mismatch between the external 50 Ω drive circuit and the travelling-wave electrode of the modulator, which has an impedance of only approx. 35 Ω, but this is only of minor effect. With optimized electrode design and doping profiles, significantly higher operating frequencies can be achieved [38,113,116]. For small-signal operation, sinusoidal drive signals are generated with an RF synthesizer with 40 GHz bandwidth. For the large-signal demonstration, an arbitrary-waveform generator (AWG) with 1 GHz bandwidth is used, allowing to generate arbitrarily pre-distorted drive signals that have a fundamental frequency of at most 100 MHz and higher harmonics of at most 500 MHz. The experiments are therefore not limited by the bandwidth of the AWG. The optical output is recorded with an optical spectrum analyzer having a resolution bandwidth of 20 MHz (Apex AP2050A).

In a first experiment, an RF synthesizer is used to generate sinusoidal drive signals with amplitudes much smaller than the π-voltage of the MZM such that the devices are operated in the linear regime. A static gate field of 0.15 V/nm is applied between the substrate and the SOI device layer to enhance the electrical bandwidth [68], which allows to reduces the drive voltage needed to obtain a given CE at high frequency shifts.

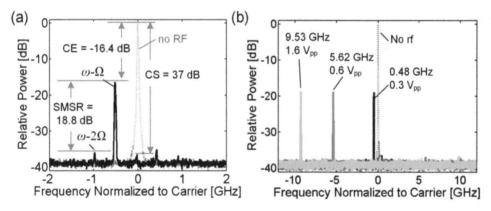

Fig. 4.8: Small-signal operation of the SOH frequency shifter. For reference, we also plot the unmodulated carrier obtained without RF signal and with the bias set for maximum transmission (dotted blue). This carrier not present in the output spectrum. (a) Modulation at $\Omega = 2\pi \times 0.475$ GHz and 0.45 V_{pp} drive voltage. The carrier suppression (CS) amounts to 37 dB, the conversion efficiency is CE = -16.38 dB and the SMSR is 18.8 dB. (b) Frequency shifting for different RF frequencies up to 10 GHz without degradation of the SMSR. The drive voltage is increased for higher frequencies to compensate for the decay of the modulation efficiency with frequency.

The optical output spectrum for a drive signal with an angular frequency of $\Omega = 2\pi \times 0.475$ GHz is depicted in Fig. 4.8(a). By fine-tuning of the bias voltage, the carrier can be suppressed by 37 dB. The shifted line at $\omega_0 - \Omega$ is clearly visible with a CE of -16.4 dB and SMSR of 18.8 dB. In the experiment, the relative phase shift between both Mach-Zehnder was chosen to be $-\pi/2$, leading to the line shifted at $\omega_0 - \Omega$ instead of $\omega_0 + \Omega$ as for a $+\pi/2$ phase shift denoted above. The sign of the relative phase shift can be chosen arbitrarily, resulting only in a mirrored output spectrum.

It has to be noted that the SMSR is limited by a strong spurious spectral line at $\omega \pm 2\Omega$ rather than at $\omega + 3\Omega$, as would be expected [22]. This is attributed to slightly different electro-optic coefficients of the two phase shifters in the arms of the two MZM, resulting in imperfect suppression of the even-order side lines. From the height of the side lines, we can estimate the ratio of the phase shifts of the MZM arms to be 1 : 0.8. This imbalance is attributed to the poling procedure: The MZM is poled by a applying a voltage via the floating ground electrodes of each MZM, while the center signal electrode is not contacted. For slots with slightly different resistivities at the glass-transition temperature, this configuration may lead to slightly different poling fields hence to slightly different electro-optic coefficients in the two arms. This can be overcome by applying a defined potential of $U_{poling}/2$ to the signal electrode during the poling process, which should lead to equal poling fields and hence equal electro-optic coefficients. Another approach to overcome the issue of an imbalance in electro-optic coefficients is based on using dual-drive configurations of the MZM, where the modulation voltage of each phase shifter can be adjusted individually. In our experiments, we demonstrate frequency shifts of up to $\Omega = 2\pi \times 10$ GHz without significant degradation of SMSR, Fig. 4.8(b). In these experiments, the drive voltage is increased along with the modulation frequency to compensate for the low-pass behavior of the device. Note that much higher frequency shifts can be achieved with SOH devices, for which small-signal modulation frequencies of 100 GHz have been demonstrated [86].

To increase the CE, a larger modulation depth and hence a larger drive amplitude are necessary. This leads to spurious lines in the optical spectrum, which need to be suppressed by using periodic drive signals which comprise harmonics of uneven order with carefully optimized amplitudes. These harmonics lead to a deformation of the drive signal, which, in combination with the nonlinear response of the MZM, generates approximately sinusoidal amplitude modulations at the output of each MZM. To experimentally verify this concept, we use an arbitrary waveform generator (AWG) with two channels to drive the device. Keeping the frequency at $\Omega = 2\pi \times 100$ MHz, the drive amplitude, waveform and electrical phase of each MZM can be controlled individually to minimize spurious side lines in the optical spectrum. Fig. 4.9(a) depicts the spectrum if purely sinusoidal signals are fed to each MZM of the frequency shifter with an amplitude a_1 of approximately the π-voltage, optimized for maximum power in the $\omega_0 - \Omega$ line. This results in a CE of -4.7 dB and in an SMSR of only 16.4 dB, in good agreement with theory and with previous results obtained from other devices that do not feature residual amplitude modulation, e.g., LiNbO$_3$ modulators, where a CE of -4.7 dB and an SMSR of 18.4 dB were reached [94].

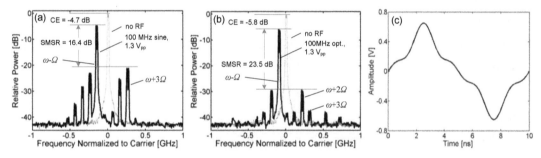

Fig. 4.9: The frequency shifter is driven with an AWG at an angular frequency of $\Omega = 2\pi \times 100$ MHz and a voltage of 1.3 V$_{pp}$ at the MZM, which is optimized for maximum conversion. Each MZM of the frequency shifter is driven by a separate channel of the AWG and amplitude, phase and shape of the waveform can hence be controlled individually. (a) Output spectrum obtained by a pure sinusoidal drive signal, resulting in an optical spectrum with good CE but small SMSR of 16.4 dB. (b) Output spectrum obtained for using a drive signal with higher harmonics, which suppress the spurious lines in the optical output spectrum. By optimizing the amplitudes of the third and the fifth harmonic, the spurious component at $\omega_0 + 3\Omega$ could be reduced by over 15 dB. The overall SMSR now amounts to 23.5 dB, limited by strong spurious components at $\omega_0 \pm 2\Omega$, which are attributed to an imbalance of the phase shifters in the MZM. This can be overcome by optimized device design. Without these components, an SMSR of 31 dB could be achieved. (c) Waveform of the optimized drive signal for one MZM, generated by
$U = a_0 \sin(\Omega t) - 0.25 \cdot a_0 \sin(3\Omega t) + 0.125 \cdot a_0 \sin(5\Omega t)$.

We then use an experimentally optimized drive signal comprising the third and fifth harmonic, $U = a_0 \sin(\Omega t) - 0.25 \cdot a_0 \sin(3\Omega t) + 0.125 \cdot a_0 \sin(5\Omega t)$, see Fig. 4.9(c) for the time-domain electrical waveform. This significantly reduces the spurious side modes in the optical spectrum, Fig. 4.9(b). The previously dominant line at $\omega_0 + 3\Omega$, which is minimized by the third harmonic of the drive signal, is now 31 dB below the line at $\omega_0 - \Omega$. Note that the SMSR is again limited by the spurious lines at $\omega_0 \pm 2\Omega$, which in theory should not exist. These lines originate from the unequal efficiency of the phase shifters in the MZM, leading to an SMSR of 23.5 dB. This can be avoided by improved device design, enabling SMSR of better than 30 dB. The CE for modulation with pre-distorted signals amounts to -5.8 dB, quite close to the theoretically predicted value of - 6 dB. The fact that the theoretically predicted CE amounts to - 6 dB can be understood intuitively: For ideal compensation, the field at the output of the MZM is perfectly sinusoidal and the intensity follows a squared sinusoidal. As a consequence, half the power (3 dB) is lost on average in each of the two MZM. Moreover, the outputs of both MZM are merged with a relative phase difference of π/2 to form the single-sideband output of the IQ modulator, thereby losing again 3 dB of power. Note that this is not in contradiction to the conversion efficiency of -4.7 dB predicted and achieved for the case of large-signal modulation with sinusoidal waveform: When driving the phase modulators of the MZM with a pure sinusoidal, the field at the output of the MZM is a distorted sinusoidal, for which the average power is larger than half the peak power. As a consequence, only 1.7 dB of power is lost in each MZM, which, together with the 3 dB of loss associated with the power combiner of the outer IQ structure, leads to a theoretical CE of -4.7 dB. In summary, we can conclude that a frequency shifter with an SMSR of better than 30 dB and a CE of approximately 6 dB can be realized on the SOH platform using the SSB configuration combined with temporal shaping of the drive signal.

5. Summary and Outlook

We have demonstrated for the first time a silicon-based single-sideband frequency shifter. The device is realized by SOH integration, enabling highly efficient pure phase modulation without any residual amplitude fluctuations. In small-signal operation, the device allows for broadband operation up to 10 GHz.

4.3 Silicon-organic hybrid frequency shifter

Using large drive amplitudes along with temporal shaping of the drive signal enables a good CE of -5.8 dB while maintaining high SMSR of more than 23 dB. The SMSR can be improved further by using optimized device designs or dual-drive configurations of the modulators.

While SOH devices show already outstanding performance, further research is needed to improve the long-term stability of the EO materials. With the relatively low glass transition temperature of the DLD164 material used here, the thermally activated reorientation of the chromophores limits the lifetime of current devices. For future device generations, the EO molecules can be modified with specific crosslinking agents. For cross-linked EO materials, stable operation above 200 °C has already been demonstrated [39,46,69]. Detailed investigations on the lifetime and stability of SOH devices are currently in progress. Moreover, there is still a series of practical device-related aspects that need systematic investigation. As an example, drift of the operating point, which is well known and thoroughly investigated in LiNbO$_3$ [117] and other devices with insulating electro-optic layer, is also observed in SOH devices and requires further investigation.

Appendix

In the following section, the equations for SSB modulation are described in more detail, including temporal shaping of the drive signal. For the analysis, we assume an ideal sinusoidal amplitude transmission function for each of the MZM, which are biased in the zero-transmission point, see Fig. 4.5. We furthermore assume that the output signals of the two MZM are combined with a relative phase shift of π/2, corresponding to the case of an ideal IQ modulator.

The spectrum at the output of the frequency shifter as depicted in Fig. 4.4(a) is the coherent addition of all separate signals after the four phase modulators. The phase shift obtained for a sinusoidal drive signal at an angular frequency Ω without predistortion can be written as

$$\begin{aligned}\Phi(t) &= a_1 \sin(\Omega t) \\ \Phi'(t) &= -b_1 \cos(\Omega t),\end{aligned} \qquad (4.22)$$

where and a_1 and b_1 represent the amplitude of the phase shift in the upper and the lower MZM, respectively. The complex optical field E at the output of the IQ modulator can then be described by

$$E = \frac{1}{4} E_0 e^{j\omega_0 t} \left(e^{ja_1 \sin(\Omega t)} - e^{-ja_1 \sin(\Omega t)} + je^{-jb_1 \cos(\Omega t)} - je^{jb_1 \cos(\Omega t)} \right) \quad (4.23)$$

where E_0 is the amplitude and ω_0 is the frequency of the optical signal at the input. Using the Jacobi-Anger expansion for expressions of the form $e^{jz\cos\theta} = \sum_{n=-\infty}^{\infty} j^n J_n(z) e^{jn\theta}$ and $e^{jz\sin\theta} = \sum_{n=-\infty}^{\infty} J_n(z) e^{jn\theta}$, where $J_n(z)$ is the n-th order Bessel function of the first kind [118], and exploiting the relation $J_n(-z) = (-1)^n J_n(z)$, Eq. (4.23) can be written as

$$E = \frac{1}{4} E_0 e^{j\omega_0 t} \left(\begin{array}{c} \sum_{n=-\infty}^{\infty} J_n(a_1) e^{jn\Omega t} - \sum_{n=-\infty}^{\infty} (-1)^n J_n(a_1) e^{-jn\Omega t} \\ +j \sum_{n=-\infty}^{\infty} (-j)^n J_n(b_1) e^{jn\Omega t} - j \sum_{n=-\infty}^{\infty} j^n J_n(b_1) e^{-jn\Omega t} \end{array} \right). \quad (4.24)$$

Condensing Eq. (4.24) for $a_1 = b_1$, the amplitude of the optical field at the output of the frequency shifter can be expressed as

$$E = E_0 \sum_{m=4n+1} J_m(a_1) e^{j(\omega_0 + m\Omega)t} \quad \text{for } n \in \mathbb{Z} \quad (4.25)$$

This corresponds to the output spectrum in Fig. 4.4(b) with lines at $\omega_0 - 3\Omega$, $\omega_0 + \Omega$, $\omega_0 + 5\Omega$ etc. For a frequency shifter, the line at $\omega_0 + \Omega$, i.e. $J_1(a_1)$, has to be maximized, which requires the amplitude a_1 of the drive signal to be close to 1.84, where the first-order Bessel function assumes a local maximum. In this regime, the higher-order Bessel function ($|m| > 7$) give no significant contribution to the output spectrum and therefore the lines at these higher frequencies can be neglected. Note that the sign of the frequency shift can be reversed by reversing the sign of the phase shift φ_Q in Fig. 4.4. Equation. (4.23) refers to $\varphi_Q = +\pi/2$; the corresponding relation for the case of $\varphi_Q = -\pi/2$ would read

$$E = \frac{1}{4} E_0 e^{j\omega_0 t} \left(e^{ja_1 \sin(\Omega t)} - e^{-ja_1 \sin(\Omega t)} - je^{-jb_1 \cos(\Omega t)} + je^{jb_1 \cos(\Omega t)} \right) \quad (4.26)$$

4.3 Silicon-organic hybrid frequency shifter

Using the same procedure as outlined above, we find that the field at the output of the device is shifted towards lower frequencies,

$$E = E_0 \sum_{m=4n+1} J_m(a_1) e^{j(\omega_0 - m\Omega)t} \quad \text{for } n \in \mathbb{Z}. \tag{4.27}$$

To minimize the amplitudes at the spurious frequencies $\omega_0 - 3\Omega$, $\omega_0 + 5\Omega$, and $\omega_0 - 7\Omega$, a pre-distortion can be applied to the drive signal by introducing higher harmonics into the waveform. Exemplarily, the compensation for one spurious frequency $\omega_0 - 3\Omega$ will be explained in the following. When adding the third harmonics to the electrical drive signals of the two MZM in the IQ modulator, Eq. (4.22) changes to

$$\begin{aligned}\Phi(t) &= a_1 \sin(\Omega t) - a_3 \sin(3\Omega t), \\ \Phi'(t) &= -b_1 \cos(\Omega t) - b_3 \cos(3\Omega t).\end{aligned} \tag{4.28}$$

In this relation, the quantities a_1, b_1 and a_3, b_3 represent the amplitudes of the phase shift in the upper and the lower MZM at frequencies Ω and 3Ω, respectively. This leads to phase shifts of the form

$$E = \frac{1}{4} E_0 e^{j\omega_0 t} \begin{pmatrix} e^{ja_1 \sin(\Omega t)} e^{-ja_3 \sin(3\Omega t)} - e^{-ja_1 \sin(\Omega t)} e^{ja_3 \sin(3\Omega t)} \\ + j e^{-jb_1 \cos(\Omega t)} e^{-jb_3 \cos(3\Omega t)} - j e^{jb_1 \cos(\Omega t)} e^{jb_3 \cos(3\Omega t)} \end{pmatrix}. \tag{4.29}$$

The Jacobi-Anger expansion can be applied to this equation, resulting in

$$E = \frac{1}{4} E_0 e^{j\omega_0 t} \begin{pmatrix} \sum_{n=-\infty}^{\infty} \sum_{k=-\infty}^{\infty} J_n(a_1) e^{jn\Omega t} (-1)^k J_k(a_3) e^{jk3\Omega t} \\ - \sum_{n=-\infty}^{\infty} \sum_{k=-\infty}^{\infty} (-1)^n J_n(a_1) e^{jn\Omega t} J_k(a_3) e^{jk3\Omega t} \\ + j \sum_{n=-\infty}^{\infty} \sum_{k=-\infty}^{\infty} (-j)^{n+k} J_n(b_1) e^{jn\Omega t} J_k(b_3) e^{jk3\Omega t} \\ - j \sum_{n=-\infty}^{\infty} \sum_{k=-\infty}^{\infty} j^{n+k} J_n(b_1) e^{jn\Omega t} J_k(b_3) e^{jk3\Omega t} \end{pmatrix}. \tag{4.30}$$

For practical applications, the modulation amplitudes a_1 and b_1 are usually smaller than 1.84, where J_1 has its maximum, and a_3 and b_3 are usually smaller than 0.5. In this regime, the dominant contributions to the four sums on the right-hand side of Eq. (4.30) arise from products of the form $J_0(a_1) J_k(a_3)$ and

$J_n(a_1) J_0(a_3)$ that contain 0th-order Bessel functions. Taking into account only these contributions, only frequency components at $\omega_0 + \Omega$, $\omega_0 - 3\Omega$, $\omega_0 + 5\Omega$ ... remain in Eq. (4.30). Assuming further $a_1 = b_1$ and $a_3 = b_3$ and sorting by frequency components, Eq. (4.30) can be simplified to

$$E = E_0 e^{j\omega_0 t} \begin{pmatrix} \ldots \\ +[\ldots + J_1(a_1)J_0(a_3) + \ldots]e^{j\Omega t} \\ +[\ldots + J_{-3}(a_1)J_0(a_3) - J_0(a_1)J_{-1}(a_3) + \ldots]e^{-j3\Omega t} \\ +[\ldots + J_5(a_1)J_0(a_3) + \ldots]e^{j5\Omega t} \\ +[\ldots + J_{-7}(a_1)J_0(a_3) + \ldots]e^{-j7\Omega t} \\ +\ldots \end{pmatrix} \quad (4.31)$$

where the dots (…) denote expressions that either contain products of two Bessel functions of non-zero order or contributions at frequency that are outside the range $[\omega_0 - 7\Omega, \omega_0 + 7\Omega]$. Equation (4.31) is equivalent to Eq. (4.21).

[end of paper]

5 System implementation of SOH devices

5.1 Operation at elevated temperatures

For the implementation of SOH modulators in future devices the stability and resilience against increased temperatures is of great importance. The silicon-on-insulator platform and especially silicon itself can withstand high temperatures with ease due to its material properties. Furthermore, decades of development of integrated electronics have led to well-engineered solutions for thermal load and dissipation on CMOS based circuits which can be translated to the SOI platform. However, organic materials have a much lower resilience to elevated temperatures. As discussed in Section 3.1.4.1, several mechanisms can lead to degradation of the electro-optic effect in poled EO materials. For non-crosslinked materials, the glass transition temperature is one key parameter for the stability at high temperatures.

Table 5.1: Selected parameters of the electro-optic materials used within this work. Refractive index n at 1550 nm, electro optic coefficient r_{33} in SOH devices, and glass transition temperature T_g. The data for DLD164 is taken from [17], for the binary chromophore material YLD124/PSLD41 from [18]. For SEO100 n and T_g are from [47], for the measurement of the in-device r_{33} see Section 5.1.1.

	DLD164	YLD124/PSLD41	SEO100
n	1.83	1.73	1.71
r_{33} [pm/V]	180	230	122
T_g [°C]	66	97	135

Table 5.1 shows a comparison of different electro-optic materials used within this work. The monolithic material DLD164 [114] and the binary chromophore organic glass (BCOG) YLD124/PSLD41 [66] are research materials engineered towards a high nonlinearity with electro optic coefficients above 200 pm/V, but the glass transition temperature is below 100 °C. The commercially available

guest-host material SEO100 [47] has a moderate electro optic coefficient of 120 pm/V, however, a significant higher glass transition temperature of close to 140 °C. These properties make SEO100 a promising material for SOH devices with high efficiency and good stability.

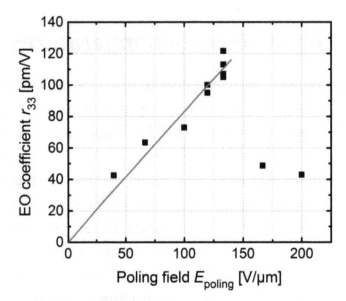

Fig. 5.1: Measured electro-optic coefficient for SOH MZM with SEO100 under different poling fields. For a poling field of 133 V/μm the maximum nonlinearity of r_{33} = 122 pm/V was reached. For poling fields above 133 V/μm the EO coefficient drops significantly, which is attributed to localized dielectric breakdown. To obtain the poling efficiency, a linear fit (red line) through the data points up to 133 V/μm is used. From the slope of the curve a poling efficiency of (0.83 ± 0.02) nm²/V² nm is extracted.

5.1.1 Static characterization of SOH devices with SEO100

SEO100 is a guest host material, consisting of amorphous propylene carbonate which is doped with 35 wt.% of a dipolar phenyltetraene chromophore [47]. Characterization of the bulk material shows an electro-optic coefficient of up to 160 pm/V at 1550 nm and after 500 h at 85 °C the material can retain 90% of its original nonlinearity [119].

To investigate the performance of SEO100 in SOH devices a study of the poling efficiency was conducted. Therefore Mach-Zehnder modulators with SEO100 as electro-optic material were used. The SOI structures were fabricated by

5.1 Operation at elevated temperatures

IMEC in a 193 nm DUV lithography. The geometrical parameters of the SOH modulator as introduced in Section 3.1.1, are as follows: the rail width is w_{rail} = 220 nm, the slab height h_{slab} = 50 nm and the slot width varies between w_{slot} = 120 nm and w_{slot} = 160 nm. To obtain the poling efficiency the devices were poled under varying poling fields E_{poling}. By measuring the π-voltage of the MZM via a DC voltage sweep, the in-device EO coefficient can be calculated according to Eq. (3.7).

The measurement results are depicted in Fig. 5.1. The highest electro-optic coefficient achieved in SOH devices with SEO100 amounts to r_{33} = 122 pm/V with an applied poling field of E_{poling} = 133 V/μm. For higher poling fields the EO coefficient is reduced significantly. This is attributed to localized dielectric breakdown at higher field strengths. The poling efficiency of the material system can be extracted from the linear dependence between poling field and EO coefficient for lower poling fields. A linear regression of the measured data, indicated in Fig. 5.1 with a red line, leads to a poling efficiency of r_{33} / E_{poling} = (0.83 ± 0.02) nm^2/V^2. This is comparable to the poling efficiency of the BCOG PSLD41/YLD124 and the monolithic chromophore DLD164 with 1.17 nm^2/V^2 and 0.93 nm^2/V^2, respectively. However, compared to SEO100, both materials can sustain higher poling fields above 200 V/μm before dielectric breakdown sets in and hence higher EO coefficients can be achieved.

Furthermore, the alignment stability at different temperatures was investigated. The poled MZM were placed on a hotplate under ambient atmosphere with a defined temperature. The electro-optic effect was measured in intervals via DC sweeps at room temperature and the sample was placed again on the hotplate. The measurement data is plotted in Fig. 5.2. Over 10 h the increase of the π-voltage was measured for temperatures of 60 °C, 80 °C and 100 °C. While for 100 °C the π-voltage almost doubles after 2 h, at 80 °C the degradation of the nonlinearity is significantly slowed after a burn-in phase of 2 h. After 10 h at 80 °C the π-voltage increased by approximately 20% and the roll-off indicates to a plateau, similar to the measurements in bulk SEO100 [119]. For 60 °C the increase in π-voltage is only 5 % over 10 h.

5 System implementation of SOH devices

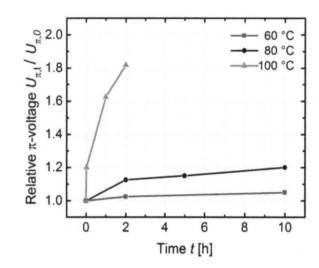

Fig. 5.2: Measurement of the increase in π-voltage over time for SOH MZM with SEO100. The devices were tested at 60 °C, 80 °C and 100 °C for 10 h. While for 100 °C a large increase is measured after the first 2 h, for 80 °C and 60 °C after a burn-in of 2 h the increase of the π-voltage is significantly reduced.

To confirm that the increase in π-voltage is related to the alignment stability of the material and not caused by photo-bleaching or other permanent damage the devices were first tempered in vacuum at the glass transition temperature without any applied poling field and subsequently the poling procedure was repeated. The devices showed the same π-voltage after re-poling as in the initial stability testing. Hence the degradation of the material can be attributed to the alignment stability of the material at elevated temperatures.

5.1.2 Signal generation with silicon-organic hybrid modulators at increased temperature

Besides the characterization of SOH devices under static conditions, the performance and stability of an SOH MZM, cladded with the commercially available EO material SEO100, was tested under high-speed modulation and elevated temperature in ambient atmosphere. This section was published in a scientific journal [16].

[start of paper]

Generation of 64 GBd 4ASK signals using a silicon-organic hybrid modulator at 80 °C

Optics Express Vol. 24, Issue 9, pp. 9389-9396 (2016)

DOI: 10.1364/OE.24.009389   © 2016 Optical Society of America

M. Lauermann,[1] S. Wolf,[1] W. Hartmann,[1] R. Palmer,[1,5]
Y. Kutuvantavida,[1] H. Zwickel,[1] A. Bielik,[2] L. Altenhain,[2] J. Lutz,[2]
R. Schmid,[2] T. Wahlbrink,[3] J. Bolten,[3] A. L. Giesecke,[3] W. Freude,[1]
and C. Koos[1,4]

[1] *Karlsruhe Institute of Technology, Institute of Photonics and Quantum Electronics, 76133 Karlsruhe, Germany*

[2] *Micram Microelectronic GmbH, 44801 Bochum, Germany*

[3] *AMO GmbH, 52074 Aachen, Germany*

[4] *Karlsruhe Institute of Technology, Institute of Microstructure Technology, 76344 Eggenstein-Leopoldshafen, Germany*

[5] *Now with: Coriant GmbH, 81541 Munich, Germany*

We demonstrate a silicon-organic hybrid (SOH) Mach-Zehnder modulator (MZM) generating four-level amplitude shift keying (4ASK) signals at a symbol rates of up to 64 GBd both at room temperature and at an elevated temperature of 80 °C. The measured line rate of 128 Gbit/s corresponds to the highest value demonstrated for silicon-based MZM so far. We report bit error ratios of 10^{-10} (64 GBd BPSK), 10^{-5} (36 GBd 4ASK), and 4×10^{-3} (64 GBd 4ASK) at room temperature. At 80 °C, the respective bit error ratios are 10^{-10}, 10^{-4}, and 1.3×10^{-2}. The high-temperature experiments were performed in regular oxygen-rich ambient atmosphere.

1. Introduction

Power-efficient photonic-electronic interfaces operating at high speed are key elements of high-performance data and telecommunication networks. With the widespread adoption of coherent transmission techniques, Mach-Zehnder modulators (MZM) and MZM-based in-phase/quadrature (IQ) modulators have become central building blocks of such interfaces. To support advanced modulation formats, MZM must be capable of reliably generating multilevel signals at high symbol rate, low energy consumption, small footprint and low cost. Silicon photonics is an especially attractive integration platform for such devices: The high refractive index contrast of silicon-on-insulator (SOI) waveguides allows for compact devices and high integration density, while mature CMOS processing enables fabless mass production and opens a path towards large-scale co-integration with electronics [8,78]. Since the intrinsic inversion symmetry of the silicon crystal lattice prohibits any technically relevant second-order nonlinearities, current high-speed silicon MZM have to rely on carrier injection or depletion in reverse-biased *p-n* junctions that are integrated into the waveguide core [7,55]. This leads to rather low modulation efficiencies with minimum voltage-length products of the order of $U_\pi L = 10$ Vmm [50] and results in comparatively large, millimeter-long modulators which still require high drive voltages of up to 5 V [11]. For simple on-off-keying (OOK), the highest data rate demonstrated with an all-silicon MZM so far amounts to 70 Gbit/s [120]. Using advanced modulation formats, the highest symbol rates amount to 56 GBd, achieved in a QPSK signaling experiment with a line rate of 112 Gbit/s per polarization at a bit error ratio of 1×10^{-2} [81]. Higher line rates of up to 180 Gbit/s per polarization have been achieved by using more complex modulation formats such as 64QAM at symbol rates of 30 GBd [13], or by using electrical OFDM [121], but these implementations need extensive and power-hungry pre- and post-processing in the electrical domain to compensate for the impairments caused by the modulators' amplitude-phase coupling. The limitations of conventional all-silicon modulators can be overcome by silicon-organic hybrid (SOH) integration, which exploits organic electro-optic (EO) materials in combination with conventional SOI slot waveguides to realize pure phase modulators that do not suffer from amplitude-phase coupling [14,32,56]. With this approach, devices with significantly reduced voltage-length products of only $U_\pi L = 0.5$ Vmm have been demonstrated, enabling modulation at record-low energy consumption of only 1.6 fJ/bit [17,18]. In transmission experiments, SOH MZM have been demonstrated to enable data rates of 100 Gbit/s using simple OOK [113]. Similarly, SOH IQ modulators are perfectly suited for signaling with higher-order modulation formats such as 16QAM [122]. Symbol rates of up to 40 GBd and line rates of up 160 Gbit/s have been demonstrated experimentally [116]. The high efficiency of SOH devices even enables direct interfacing of IQ modulators to the binary outputs of an FPGA to generate 16QAM data streams without using preamplifiers or digital-to-analog converters [123]. However, while SOH modulators show great promise to further increase both symbol and data rates, the wider use of these devices is still hampered by doubts as to the temperature stability of the organic materials, in particular regarding the decay of acentric order in the poled material when operated at elevates temperatures.

In this paper we demonstrate that SOH Mach-Zehnder modulators can generate multi-level signals at symbol rates of up to 64 GBd, both at room temperature and at elevated temperatures of 80 °C. The high-temperature experiments were performed in regular oxygen-rich ambient atmosphere using an EO material, which is specified to be stable up a temperature of 85 °C [119]. In the experiment, we generate both binary phase shift keying (BPSK) and 4-level amplitude shift keying (4ASK) signals with a single 750 µm long SOH Mach-Zehnder modulator [124], achieving line rates of up to 128 Gbit/s. Our experiments represent the first demonstration of SOH devices operating at an elevated temperature, while achieving the highest data rate so far demonstrated for a silicon-based MZM.

2. Silicon-Organic Hybrid Modulator

Silicon-organic hybrid modulators rely on the interaction of the light guided by an SOI waveguide with an EO cladding material which is exposed to an externally applied electric field [56]. Fig. 5.3(a) shows a schematic (top) and a cross section of the SOH Mach-Zehnder modulator (MZM): Standard SOI strip waveguides form the interferometer together with multimode interference couplers (MMI) operating as amplitude splitter and combiner. In each arm, the phase modulators are realized by SOI slot waveguides, the cross-section and the profile of the dominant optical electric field amplitude of which are shown in Fig. 5.3(b). The silicon rails and the slot have a width of w_{rail} = 240 nm and w_{slot} = 120 nm, respectively. The rails are connected to a coplanar ground-signal-ground (GSG) transmission line, Fig. 5.3(a), via conductive, slightly n-doped silicon slabs with a height of h_{slab} = 70 nm, Fig. 5.3(c).

5 System implementation of SOH devices

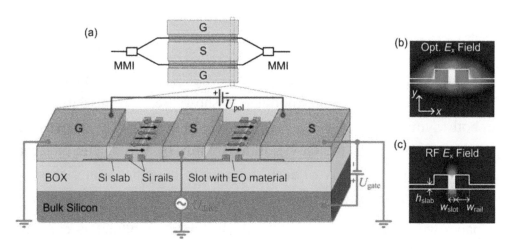

Fig. 5.3: Silicon-organic hybrid modulator. (a) Schematic of and cross-section of a silicon-organic hybrid (SOH) Mach-Zehnder modulator (MZM). Standard silicon strip waveguides and multimode interference coupler (MMI) form the waveguide interferometer. The phase shifter in each arm consists of a slot waveguide (rail width $w_{rail} = 240$ nm, slot width $w_{slot} = 120$ nm). The waveguides are electrically connected to a ground-signal-ground (GSG) RF transmission line via n-doped silicon slabs (thickness $h_{slab} = 70$ nm). A gate voltage U_{Gate} between the Si substrate and the SOI device layer improves the conductivity of the silicon slab and hence the bandwidth of the device. As cladding material the commercially available electro-optic (EO) organic cladding material SEO100 from Soluxra, specified for an operation temperature of up to 85 °C, is deposited on the chip via spin coating. To align the chromophores a poling process is performed. Therefore a voltage U_{pol} is applied via the floating ground electrodes of the device while heated to the glass temperature of the EO material. This aligns the chromophores in the slots of both waveguides (black arrows). After cooling down, the orientation of the chromophores is fixed. A modulating RF field (red arrows) can now be applied to the GSG line, resulting in a push-pull operation of the modulator. (b) E_x component of the optical field in the slot waveguide. (c) Dominant E_x component of the electrical drive signal below the RC limit. Due to the doping of the slabs the applied voltage drops predominantly across the slot, leading to a strong confined of both fields in the slot and hence efficient modulation.

For the fundamental quasi-TE mode, the high refractive-index contrast between waveguide and EO cladding leads to an enhancement of the optical field within the slot, while a voltage applied to the transmission line drops mainly across the narrow slot leading a high modulation radio frequency (RF) field. Both fields overlap strongly, leading to an efficient modulation, see Fig. 5.3(b) for a contour plot of the dominant component of the optical mode ($E_{x,opt}$) and Fig. 5.3(c) for a plot of the $E_{x,RF}$ component of the modulation field. Due to the high modulation efficiency, the MZM can be as short as 750 µm. The transmission line of the modulator is designed for a characteristic impedance of 50 Ω, matched [38,122] to the characteristic impedance of common RF equipment. The bandwidth of the SOH device is typically limited by the RC time constant formed by the slot capacitance and the finite conductivity of the resistive slabs. To increase the slab conductivity, a gate voltage U_{gate} can be applied between the silicon substrate and the device layer to accumulate electrons at the interface between Si slabs and buried oxide (BOX), see Fig. 5.3(a) [68,86]. This decreases the RC time constant of the device and increases the EO 3 dB bandwidth to more than 32 GHz, the Nyquist frequency of the 64 GBd data signal. This values was estimated by simulations and by measurements with similar devices. We may hence conclude that the signal quality in our transmission experiment was mainly limited by the 30 GHz drive amplifier, and not by the modulator itself. Without gate voltage, the EO bandwidth falls below 10 GHz. Optimized doping profiles allow to omit the gate voltage for future device generations, and we recently demonstrated an SOH device operating without gate voltage at 80 Gbit/s OOK [113]. For fabrication, we use a standard SOI wafer with a 220 nm thick device layer and a 3 µm thick BOX. The silicon waveguides are structured using electron beam lithography, while optical lithography is employed for defining the metallization. The chip is clad with the commercially available electro-optic material SEO100 from Soluxra, which is specified for operation at temperatures up to 85 °C [119]. The cladding material is poled by heating it close to its glass transition temperature of 140 °C while applying a poling voltage across the two floating ground electrodes of the MZM. Half of the voltage drops across each slot, resulting in an ordered orientation of the dipolar chromophores in the slot as indicated by the black arrows in Fig. 5.3(a). After cooling the device to room temperature, the orientation of the chromophores freezes and the poling voltage source is removed. The red arrows in Fig. 5.3(a) indicate the RF field applied to the GSG electrodes. It is antisymmetric with respect to the orientation of the chromophores, resulting in opposite phase shifts in the two arms of the MZM. Hence push-pull operation is achieved by connecting a single-ended data signal to the transmission line.

5 System implementation of SOH devices

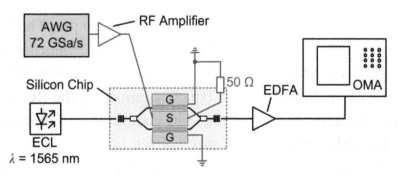

Fig. 5.4: Schematic of the experimental setup. Electrical multilevel drive signals at symbol rates 36 GBd and 64 GBd are generated by an arbitrary waveform generator (AWG) with up to 72 GSa/s from Micram Instruments, AWG6020. The DC bias voltage for adjusting the operating point is applied via a bias-T (not drawn) to the MZM and the transmission line is terminated with 50 Ohm to avoid back reflections. An external-cavity laser (ECL) at 1565 nm is used as an optical source and coupled to the MZM chip via fibers and grating couplers. The optical output signal is amplified by an EDFA and subsequently fed into an optical modulation analyzer (OMA) in homodyne configuration for detection.

3. Experimental Setup

The experimental setup for signal generation is sketched in Fig. 5.4. An arbitrary waveform generator (AWG, Micram Instruments model AWG6020) generates the electrical NRZ drive signal in form of a pseudo-random bit sequence with a length of $2^{11}-1$. An electrical amplifier with 30 GHz bandwidth amplifies the signal to a peak-to-peak voltage of 2 V. This modulation voltage is coupled to the GSG transmission line via microwave probes. An external cavity laser (ECL) generates the optical carrier, which is coupled to the chip via grating couplers (GC). The optical output of the modulator is coupled to an erbium-doped fiber amplifier (EDFA) for compensating the excess loss of 21 dB which is dominated by the loss of the non-optimized grating couplers of 6 dB per facet. The MZM features an insertion loss of 9 dB, which is attributed to losses in passive components such as MMI and strip-to-slot converters (< 1 dB in total) [71], to propagation loss in the access waveguides caused by sidewall roughness (1 dB), and to free-carrier absorption [49] ($\approx$ 1 dB) as well as scattering loss (6 dB) in the 750 µm long MZM slot waveguides. It is expected that these losses can be greatly reduced in future devices: Slot waveguides can be fabricated with propagation losses as low as 0.65 dB/mm [125], and optimizing the doping profile can reduce the doping-related losses to below 1 dB/mm. At the receiver side, an optical modulation analyzer (OMA) Agilent N4391A is used for homodyne detection and signal analysis. The OMA performs standard digital post-processing comprising polarization demultiplexing, phase estimation, and channel equalization. For signals with a symbol rate of 36 GBd the drive voltage at the input of the modulator is measured to be 2 V_{pp}.

For 64 GBd signals we use a linear electrical pre-emphasis to compensate low-pass characteristics of the setup, which are predominantly caused by the 30 GHz driver amplifier. The necessary pre-equalizer parameters are estimated at the receiver side and then applied to the AWG. Pre-equalization includes an increased voltage swing at the output of the AWG for compensating the low-pass roll-off. The resulting voltage level at the input of the modulator is estimated to remain at 2 V_{pp}. A constant gate field of 0.1 V/nm is applied between the silicon substrate and the device layer to improve the bandwidth of the device. The silicon chip with the SOH modulator is placed on a metallic sample holder which can be heated via a thermoelectric heater. The sample temperature is measured by a resistive temperature detector inside the sample holder and close to the silicon chip. Measurements are performed at room temperature as well as at 80 °C, without taking any measures to shield the devices from ambient oxygen. As modulation format, we use BPSK and 4ASK at symbol rates between 36 GBd and 64 GBd. Depending on the symbol rate, the AWG is operated at sampling rates of 72 GSa/s and 64 GSa/s, respectively.

4. Measurement

Measured eye patterns and constellation diagrams are shown in Fig. 5.5. The first set of measurements is performed at room temperature, upper row. For a second set, the sample is operated as before, but heated to 80 °C, lower row. After one hour at 80 °C in ambient atmosphere, we start the data recording. As a quantitative measure of the signal quality, we use the measured bit-error ratio (BER) and the error vector magnitude (EVM_m), which describes the effective distance of a received complex symbol from its ideal position in the constellation diagram, using the maximum length of an ideal constellation vector for normalization [77]. For operation of the device at 36 GBd, we find an average π-voltage of $U_\pi = 2.2$ V, estimated from the compression of the outer points in the 4ASK constellation diagrams. The corresponding voltage-length product at high-speed operation is therefore $U_\pi L = 1.6$ Vmm, which leads to an estimated EO coefficient of $r_{33} = 60$ pm/V. This $U_\pi L$-product is slightly higher than the values measured for best-in-class devices [17,18], which we attribute to imperfect poling and to the fact that the material used in this work is not optimized only for highest EO activity but also for increased stability. There is vast potential in improving device performance by using cross-linkable materials with EO coefficients beyond 100 pm/V [48]. Nevertheless, even in the current experiment, the measured $U_\pi L$-product is well below the 10 Vmm that are to be expected for conventional depletion-type devices [50].

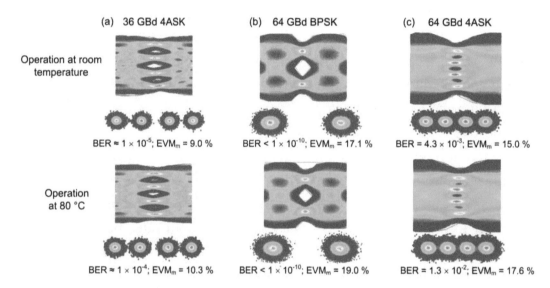

Fig. 5.5: Evaluation of the signal generation experiment. The top row depicts the measured constellation diagrams and eye diagrams for operation at room temperature. In the bottom row the corresponding data for operation at 80 °C is shown. (a) For signaling at 36 GBd with 4ASK (72 Gbit/s) at room temperature we measure an EVM_m of 9.0 %, and the BER is estimated to be 1×10^{-5}, too few errors were measured for a statistical significance. At 80 °C the EVM_m is with 10.3 % slightly increased and the BER is estimated to be 1×10^{-4}. No pre-emphasis is used for 36 GBd signaling. The peak-to-peak voltage at the modulator is measured to be 2 V_{pp}. (b) BPSK signaling at 64 GBd shows error free performance at room temperature and 80 °C. No errors could be measured within our record length of 62.5 μs and the EVM_m indicates a BER $< 10^{-10}$. (c) For 4ASK signals at 64 GBd (128 Gbit/s) we measure an EVM_m of 15.0 % and a BER of 4.3×10^{-3} at room temperature. At 80 °C the EVM_m is 17.6 % and the measured BER is 1.3×10^{-2}. For 64 GBd signals an electrical pre-emphasis was used to compensate for the low-pass characteristics in the setup. The increased voltage swing at the output of the AWG compensates for the roll off at high frequencies and the drive voltage level is estimated to remain at 2 V_{pp}.

For 36 GBd 4ASK signaling at room temperature, the EVM_m is measured to be 9.0 %, and the BER is estimated to be approximately 1×10^{-5}. Too few errors were recorded for directly determining a statistically significant BER value. At 80 °C, the EVM_m increases slightly to 10.3 %, and the BER is estimated to be 1×10^{-4}. The corresponding eye diagrams and the constellation diagrams are depicted in Fig. 5.5(a). For both temperatures, the BER are well below the threshold of 4.5×10^{-3} for hard-decision forward error correction (FEC) with 7% overhead [88]. At a symbol rate of 64 GBd we use a pre-emphasis to compensate for the low-pass characteristics in the measurement setup, which predominantly arise from the 30 GHz preamplifier (model SHF 807). The results for BPSK signaling at 64 GBd are shown in Fig. 5.5(b). The signal is error-free both for room temperature and for operation at 80 °C, i.e., no errors could be measured within our record length of 62.5 µs (8×10^6 bits). The EVM_m values of 17.1 % and 19.0 % indicate a BER smaller than 10^{-10} [77]. Fig. 5.5(c) depicts the data for 64 GBd 4ASK signals. At room temperature, the measured EVM_m is 14.7 % and the measured BER amounts to 4.3×10^{-3}, just below the threshold for hard-decision FEC. For operation at 80 °C, the EVM_m is measured to be 17.6 %, and the measured BER is 1.3×10^{-2}, still below the threshold of 2.4×10^{-2} for soft-decision FEC with 20 % overhead [89]. At room temperature, the line rate (net data rate) of the 64 GBd 4ASK signal amounts to 128 Gbit/s (120 Gbit/s). These figures represent the highest values demonstrated so far for a silicon-based Mach-Zehnder modulator.

The slightly reduced signal quality at elevated temperatures is attributed to an increased π-voltage after burn-in. Using again the compression of the outer points in the 4ASK constellation diagrams to extract the π-voltage, we estimate a 12% increase as an upper limit for the degradation after burn-in. Apart from the burn-in phase, the performance of our modulator stayed constant over the entire measurement. Moreover, we did not observe any significant degradation of the devices over several months of repeated use and storage under normal laboratory conditions at room temperature. This is well in line with previous reports on the SEO100 material, where, after a short burn-in period, more than 90% of the material efficiency was retained for more than 500 hours [119]. These experiments provide a first proof-of-principle that SOH modulators can indeed be operated at elevated temperatures. A detailed investigation of aging and temperature stability of organic EO materials is subject to ongoing research. By using modified cross-linkable chromophores, the alignment stability of the chromophores can be further improved [39]. The viability of this approach has been demonstrated for EO compounds with high r_{33}, where material stability of up to 200°C has been achieved [48,46].

5. Summary

We demonstrate the first high-speed operation of an SOH modulator at an elevated temperature of 80 °C in regular oxygen-rich ambient atmosphere. A single 750 µm long SOH Mach-Zehnder modulator is used to generate multilevel signals at symbol rates of up to 64 GBd. The achieved line rate of 128 Gbit/s and the corresponding net data rate of 120 Gbit/s correspond to the highest value demonstrated for a silicon-based Mach-Zehnder Modulator so far. Based on these findings and on progress in EO material research, we expect that SOH modulators evolve towards high-performance devices with high reliability.

[end of paper]

5.2 Fabrication of SOH devices on an industrial silicon-photonic platform

As outlined in Section 1, silicon photonic integration and the SOI platform is a powerful technology which allows to combine various optical elements on one photonic integrated circuit (PIC) and the fabrication can rely on mature CMOS based processes. SOH integration is based on the SOI platform but has nevertheless distinct requirements on device geometries and the process flow. The SOH devices discussed in the previous sections where fabricated using either electro-beam writing for the lithography of the silicon structures or a highly customized 193 nm DUV process. In the following, the fabrication of SOH devices on a commercial SOI process with 248 nm lithography is discussed. Using such a standard process allows the co-integration of SOI devices with a large variety of photonic elements which are typically available via foundry services [8]. Within this work, SOH devices were fabricated in a commercially available silicon photonics foundry run at the Institute of Microelectronics

A*Star (IME) in Singapore. The process flow comprises around 20 mask layers defining various structures such as different silicon etch depths, Germanium growth, two layer metallization and different doping levels. While for customized processes each flow step and process boundaries can be explored individually to match the desired design, there are fixed design rules for commercial process flows, ensuring the consistent and deterministic results for a wide variety of designs.

5.2.1 Special requirements of SOH devices and their implementation in a commercial process flow

The minimum feature size for 248 nm DUV lithography is typically around 200 nm, ideally suited for standard SOI devices with minimum feature sizes between 300 nm and 500 nm in the waveguide layer. Slot waveguides for SOH devices, however, have typically rail widths between 200 nm and 300 nm and a slot with around 140 nm, as discussed in Section 3.1. While the resolution of the lithography is sufficient for the rails of the waveguide it is too coarse for the definition of the slot. To realize slot widths below 200 nm an additional mask step was used within the waveguide definition at IME. While the lithographic dimensions are kept above the limitation of 200 nm, the opening in the physical hard mask could be reduced by appropriate processing. With this mask layer, slot widths of 120 nm, 140 nm and 160 nm could be realized with the 248 nm lithography, ensuring a high field confinement and therefore a high modulation efficiency of the SOH devices. Since the slot is defined with a separate mask layer and not with a self-aligned process as described in Section 3.3, precise alignment between the silicon waveguide layer and the slot layer is crucial for good operation. The alignment tolerances for the two mask layer are typically in the order of ± 50 nm. The SOH devices can tolerate the resulting asymmetry of the silicon rails, however, to avoid abrupt changes of the waveguide width, tapered designs are developed at the interface of the strip and slot waveguide to mitigate the influence of the alignment uncertainty. The thin silicon slab and the rails follow a logarithmic taper for mode conversion [71]. In order to mitigate the influence of small alignment uncertainties during the lithography, the slot layer and the layer for the silicon slab both have now an overlap with the waveguide layer.

The backend processing of a standard SOI flow comprises typically the deposition of silicon oxide which serves as waveguide cladding and is also necessary to realize the different metal layers which are connected to the silicon with electrical vias. Since SOH devices rely on the combination of SOI slot waveguides and organic electro-optic material, the oxide above the slot waveguide has to be removed in order to fill the slot with the organic material. The process flow at IME comprises a step for backend oxide removal. However, this step puts additional constraints on the design: the oxide opening has a minimum width of several µm and since it is etched through the complete backend stack, there must be in addition a safety margin from the edge of the oxide opening before any metal can be placed.

As discussed in Section 3.1 the bandwidth of a resistively coupled SOH modulator is dependent on the series resistance between the metal electrodes and the slot capacity. With the oxide opening, the metal electrodes cannot be placed as close as necessary for high speed operation. To provide still a connection between slot and metal electrodes with low resistivity, highly doped 220 nm is used. For good confinement of the optical mode the thin silicon slab is placed in the vicinity of the optical mode. See Fig. 5.6 for a cross section of the SOH structure.

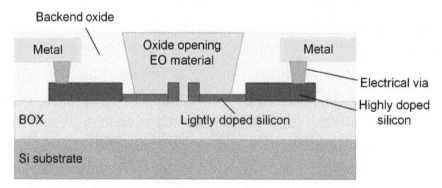

Fig. 5.6: Partial cross-section of an SOH phase modulator in the IME process flow. An oxide opening process is used to access the slot waveguide after backend processing and deposit the EO material in the slot. Highly doped silicon with a height of 220 nm is used to provide a path with low resistivity between the slot and the metal electrodes. Only in the vicinity of the slot thin, lightly doped silicon slabs are used for a good confinement of the optical mode.

5.2 Fabrication of SOH devices on an industrial silicon-photonic platform

For the travelling wave electrode, the first metal layer, closest to the silicon layer is utilized, while the second layer is used for the contact and the probe pads. This ensures, that the electrical modulation field is guided as close as possible to the optical field and reduces parasitic effects of the electrical vias. Using the highly doped 220 nm silicon, the travelling wave electrodes can be designed similar to the devices in Section 3, with an impedance close to 50 Ω.

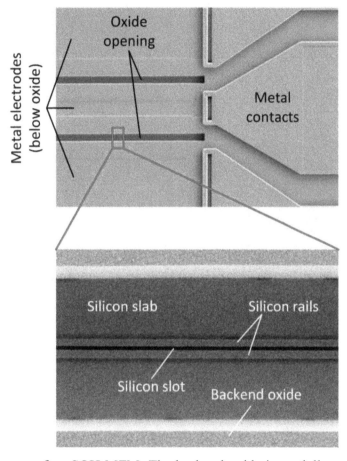

Fig. 5.7: SEM images of an SOH MZM. The backend oxide is partially removed, exposing the slot waveguide. Although two mask layers are used to define the slot waveguide, it can be seen in the SEM image that the slot is well centered within the waveguide.

5.2.2 Characterization of the fabricated devices

After fabrication of the silicon structures at IME, the devices were prepared for deposition of the organic material. To remove any residual contaminations in the slot, an additional cleaning step in a cleanroom environment was performed

5 System implementation of SOH devices

prior to deposition of the organic material. Fig. 5.7 shows image details of an SOH MZM fabricated at IME. The opening of the backend oxide above the rails is clearly visible, the metal transmission line is buried below the backend oxide. In the zoom-in the slot waveguide can be seen, the silicon slab extends below the backend oxide and the slot itself is well centered in the waveguide.

The organic material SEO100, which is described in Section 5.1.1 was chosen as organic cladding material. The material was deposited on the SOI structures via spin coating from dibromomethane. The material fills the silicon slot without any voids, and a layer thickness above 1 µm ensures that the optical mode is properly guided within the cladding. The SOH Mach-Zehnder devices are first characterized with a passive transmission measurement. An intentional imbalance in the Mach-Zehnder interferometer structures allows to record the extinction ratio as well as the insertion loss of the devices with a simple sweep over the wavelength. The fiber-to-fiber loss of a 1.1 mm long MZM is measured to be 21 dB with an extinction ratio of over 20 dB of the interferometric structure. The loss of a 0.5 mm long MZM is measured to be 15 dB also with over 20 dB extinction ratio.

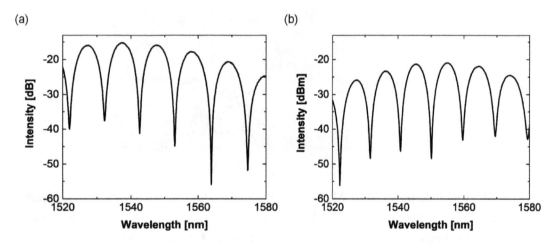

Fig. 5.8: Fiber-to-fiber transmission for MZM coated with SEO100. (a) optical transmission measurement of a MZM with a 1.1 mm long active slot waveguide section. (b) optical transmission measurement of a 0.5 mm long MZM. An intentional imbalance in the interferometer leads to the fringes with approx. 5 nm periodicity. The envelope of the transmission curves is due to the response of the grating coupler over wavelength.

5.2 Fabrication of SOH devices on an industrial silicon-photonic platform

The transmission of both MZM is depicted in Fig. 5.8, the envelope of the curve is due to the response of the grating coupler over wavelength. From test structures, the losses of the grating coupler can be estimated to be 4.5 dB per coupler, leading to on-chip losses of 12 dB for the 1.1 mm long MZM and 6 dB for the 0.5 mm long MZM.

To evaluate the electro-optic response, poling experiments were performed. Following the parameters determined in Section 5.1.1 a π-voltage of 0.9 V could be obtained in a 1.1 mm long MZM with 160 nm slot width. This corresponds to a voltage-length product of 1 Vmm and an in-device electro-optic coefficient of 140 pm/V. Besides the evaluation of the SOH devices under static conditions, a dynamic electro-optic measurement was performed to evaluate the performance under high-frequency excitation. Therefore, a vector network analyzer with calibrated photodiode was used to measure the electro-optic response over a wide frequency range. A MZM with 0.4 mm long slot waveguides was measured.

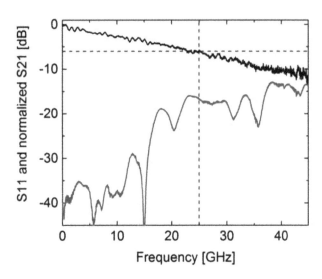

Fig. 5.9: EOE frequency response of a 0.4 mm long MZM. In black the normalized S21 parameter is depicted. The 6 dB point of the EOE response corresponds to the 3 dB point of the EO response and is indicated by a dashed line. The bandwidth of the device is measured to be 25 GHz. The S11 stays below 20 dB up to 20 GHz and below 15 dB up to 35 GHz, an indication for a good impedance matching to the 50 Ohm measurement system.

5 System implementation of SOH devices

The S21 and S11 parameter are plotted in Fig. 5.9, the 6 dB electro-optic-electric (EOE) point corresponds to the 3 dB electro-optic (EO) bandwidth and amounts for this device to 25 GHz. The relation between EO and EOE response is explained in Appendix C. The S11 parameter is below 20 dB for frequencies up to 20 GHz and below 15 dB up to 35 GHz. This indicates a proper matching of the transmission line to the 50 Ohm measurement system.

The performance of those SOH devices makes them ideal candidates for future experiments for high-speed modulation with advanced modulation formats and extended reliability studies. Furthermore, the integration of the SOH devices in the process flow of a commercially available silicon photonics foundry platform allows the use of SOH modulators for a variety of different experiments and system concepts.

6 Summary and outlook

6.1 Summary

Optical integration on the silicon photonic platform has the potential to enable the next generation of communication systems as well as new applications in sensing or metrology. The work in this thesis is focused on the investigation and development of silicon-organic hybrid IQ modulators to reduce the power consumption, increase the speed and enable the integration of such modulators in standard silicon photonic PICs.

SOH modulator for advanced modulation formats with low power consumption

By combining highly efficient EO materials with an SOH modulator in IQ configuration, a device capable of 16QAM modulation at 28 GBd with an energy consumption of only 19 fJ/bit was demonstrated. The high nonlinearity of the organic EO material PSLD124/YLD124 together with narrow SOI slot waveguides ensures operation with drive voltages as low as 0.6 V_{pp}. Using these low voltages, a drive amplifier can be omitted which helps to lower the energy consumption of a potential system even further. By directly connecting the DAC to the modulator best signal quality is ensured.

SOH modulator for highest data rates

While maintaining low drive voltages, signal generation with advanced modulation formats at highest data rates was demonstrated. The low drive voltages facilitate high-speed operation, since RF electronics with large voltage swing at high frequencies are difficult to realize and have a high power consumption. Using an SOH IQ modulator with a matched 50 Ohm transmission line as RF electrodes, 16QAM modulation with symbol rates up to 40 GBd was demonstrated. Using an SOH MZM, generation of 4ASK signals with a record high symbol rate of 64 GBd could be shown. To further extend the bandwidth of SOH devices and overcome the limitations of conventional resistively coupled slot waveguides, a new concept of capacitively coupled SOH devices is introduced and theoretically investigated. By using materials with high permittivity to couple the metallic electrodes to the slot waveguide, a large overlap of the optical field and the RF field within the slot waveguide is guaranteed. At the same time the resistive silicon slab can be omitted and therefore the associated RC time constant is not limiting the high-speed performance of future SOH devices anymore.

Frequency shifter based on SOH integration

The benefits of SOH modulators in IQ configuration are not limited to applications in optical communications. Integrated frequency shifter are important building blocks for example in vibrometry or distance metrology. Leveraging the pure phase modulation, low drive voltage, and high bandwidth of SOH modulators a single-sideband frequency shifter was demonstrated on the silicon platform. This device is capable of achieving frequency shifts up to 10 GHz, magnitudes higher than conventional implementations in silicon using thermo-optic phase shifters. To overcome the trade-off for single-sideband modulation between conversion efficiency and side mode suppression ratio, a concept for temporal shaping of the drive signal was developed and demonstrated. By adding additional harmonic frequencies in the electrical drive signal, spurious side modes can be suppressed. This results in an experimentally demonstrated conversion efficiency of -5.8 dB and a side mode suppression ratio above 23 dB.

Towards system implementations of SOH devices

SOH devices are showing great performance in experimental demonstrations. But to leverage all advantages of the silicon photonics platform it is important that the fabrication can rely on standard industrial processes. The slot waveguide as critical feature was so far typically realized with e-beam lithography or customized 193 nm DUV lithography. During this work a design and layout of SOH modulators was developed, which can be fabricated in a standard 248 nm DUV lithography process within a commercial silicon photonics foundry run. The resulting devices did show similar or even better performance than earlier devices using customized processes.

Furthermore, the commercial EO organic material SEO100 was introduced on the silicon-organic hybrid platform and a poling process for the material was developed. This material has a higher temperature resilience, which allowed the generation of 4ASK signals at symbol rates of 64 GBd, while keeping the SOH device at an elevated temperature of 80 °C, a first step to SOH devices with high performance and high reliability.

6.2 Outlook and future work

Within this work advanced SOH modulators as building blocks for complex silicon-photonic systems were demonstrated. Towards the realization of highly integrated systems co-integrated with electronic circuitry, still some challenges have to be faced. While the demonstration of SOH operation at 80 °C shows the feasibility of such an approach for practical implementations, the long-term stability of SOH devices has yet to be evaluated. Besides the organic materials used in this work, which rely on embedding the chromophores in an amorphous matrix material, new approaches for EO materials have been developed, relying on cross-linking for post-poling lattice hardening. If those materials can be used within SOH devices, stable and reliable SOH devices for operation above 100 °C are well within reach. Co-integration of SOH devices directly with the electronics opens up new possibilities for design and operation. While common RF equipment in the laboratory is all matched to 50 Ohm, a 50 Ohm system might not be the ideal source for driving SOH devices. However, integrated electronics are not limited by this quasi-standard impedance. Optimizing both as a pair, the electrical driver circuitry together with the SOH modulators, is a very promising path to significantly increase the performance and reduce the energy consumption of the whole system even further.

Appendix

A Mathematical relations

A.1 Fourier transformation

Using the angular frequency ω, the Fourier transformation $\mathcal{F}$ and its inverse of the field $\vec{E}$ is defined as follows:

$$\vec{E}(\vec{r},\omega) = \mathcal{F}\{\vec{E}(\vec{r},t)\} = \int_{-\infty}^{\infty} \vec{E}(\vec{r},t) e^{-j\omega t} dt \qquad (A.1)$$

$$\vec{E}(\vec{r},t) = \mathcal{F}^{-1}\{\vec{E}(\vec{r},\omega)\} = \frac{1}{2\pi} \int_{-\infty}^{\infty} \vec{E}(\vec{r},\omega) e^{j\omega t} d\omega \qquad (A.2)$$

A.2 Hilbert transformation

The Hilbert transformation $\mathcal{H}$ and its inverse connects the real part $s(t)$ and the imaginary part $\hat{s}(t)$ of a complex analytical signal $\underline{s}(t)$. Following the derivation in [126] it is defined as

$$\hat{s}(\tau) = \mathcal{H}\{s(t)\} = \frac{1}{\pi} \text{p.v.} \int_{-\infty}^{\infty} \frac{s(t) dt}{\tau - t}, \qquad (A.3)$$

$$s(t) = \mathcal{H}^{-1}\{\hat{s}(\tau)\} = -\frac{1}{\pi} \text{p.v.} \int_{-\infty}^{\infty} \frac{\hat{s}(\tau) d\tau}{t - \tau}, \qquad (A.4)$$

with p.v. denoting the Cauchy principal value.

B Electrical transmission lines

B.1 Lumped element circuit model of the electrical transmission line

In cases where the wavelength of the electrical wave is in the order of or smaller than the physical dimensions of the transmission line, the representation as distributed-parameter network can be used to model the transmission line. Therefore the transmission line is divided in infinitesimal small elements for which the current and voltage does not vary significantly, so a lumped element approach can be used to describe each element. The analysis in this chapter follows the derivation in [127], where more details on electrical transmission lines can be found.

An infinitesimal small transmission line element of length Δz can be described by a two-port circuit consisting of lumped elements, see Fig. B.1.

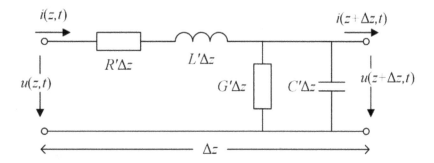

Fig. B.1: Equivalent circuit of a transmission line element of length Δz. A lumped element model is used to describe the transmission line and the evolution of the current $i(z,t)$ and voltage $u(z,t)$.

The transmission line model in Fig. B.1 is characterized by the distributed parameters R', L', G', and C', which are defined as in [127]:

Appendix

R': series resistance per length in Ω/m, for both conductors

L': series inductance per length in H/m, for both conductors

G': shunt conductance per length in S/m

C': shunt capacitance per length in F/m

Using above definitions and using Kirchoffs law, the evolution of current and voltage in time domain can be written as

$$\frac{\partial u(z,t)}{\partial t} = -R'i(z,t) - L'\frac{\partial i(z,t)}{\partial t}, \tag{B.1}$$

$$\frac{\partial i(z,t)}{\partial t} = -G'u(z,t) - C'\frac{\partial u(z,t)}{\partial t}. \tag{B.2}$$

These differential equations are also known as telegrapher's equation. Using sinusoidal signals, the equations for steady state can be simplified to

$$\frac{\mathrm{d}u(z)}{\mathrm{d}z} = -(R' + \mathrm{j}\omega L')i(z), \tag{B.3}$$

$$\frac{\mathrm{d}i(z)}{\mathrm{d}z} = -(G' + \mathrm{j}\omega C')u(z). \tag{B.4}$$

Solving Eq. (B.3) and Eq. (B.4), the wave equations

$$\frac{\mathrm{d}^2 u(z)}{\mathrm{d}z^2} - \overline{\beta}^2 u(z) = 0 \quad \text{and} \quad \frac{\mathrm{d}^2 i(z)}{\mathrm{d}z^2} - \overline{\beta}^2 i(z) = 0 \tag{B.5}$$

are obtained, with the complex propagation constant

$$\overline{\beta} = \sqrt{(R' + \mathrm{j}\omega L')(G' + \mathrm{j}\omega C')}. \tag{B.6}$$

Using the solutions

$$u(z) = u_0^+ \mathrm{e}^{-\overline{\beta}z} + u_0^- \mathrm{e}^{\overline{\beta}z} \quad \text{and} \quad i(z) = i_0^+ \mathrm{e}^{-\overline{\beta}z} + i_0^- \mathrm{e}^{\overline{\beta}z} \tag{B.7}$$

for wave propagation in positive z-direction ($\mathrm{e}^{-\overline{\beta}z}$) and negative z-direction ($\mathrm{e}^{\overline{\beta}z}$), the characteristic impedance Z can be defined as:

$$Z = \frac{u_0^+}{i_0^+} = -\frac{u_0^-}{i_0^-} = \sqrt{\frac{R' + \mathrm{j}\omega L'}{G' + \mathrm{j}\omega C'}}. \tag{B.8}$$

From Eq. (B.6) we see that the losses in the transmission line are defined by the resistance R' and conductance G', the ohmic losses of the conductor are related to R', while the dielectric losses are described by G' [127]. For a lossless transmission line, the propagation constant can be written as $\overline{\beta} = j\omega\sqrt{L'C'}$ and the characteristic impedance becomes a real number $Z = \sqrt{L'/C'}$.

B.2 Impedance matching and reflections

As described in the previous section, the voltage and current on a transmission line are related by the characteristic impedance Z. However, when the transmission line is terminated with a load, the voltage and current at the load are defined by the load impedance Z_L. Using the total voltage and current on the transmission line as defined in Eq. (B.7) a reflection coefficient Γ_R can be defined as amplitude of the reflected wave divided by the amplitude of the incident wave [127]

$$\Gamma_R = \frac{u_0^-}{u_0^+} = \frac{Z_L - Z}{Z_L + Z}. \tag{B.9}$$

For a matched load where $Z_L = Z$, no reflections occur and $\Gamma_R = 0$. For $Z_L \neq Z$ a reflection occurs at the load and a standing wave is formed on the transmission line. If the impedance of the load is larger than the impedance of the line, Γ_R is positive, i.e. the phase of the reflected wave u_0^- directly at the load is the same as the phase of the incident wave u_0^+, both interfere constructively at the position of the load, hence a antinode of the standing wave is located at the load. When the impedance of the load is smaller than the line impedance, Γ_R is negative and the phase of the reflected wave is flipped by 180°. Therefore a node is located at the position of the load.

B.3 Lossy transmission lines

The complex propagation constant $\overline{\beta}$ from Eq. (B.6) can be rewritten as

$$\overline{\beta} = j\omega\sqrt{L'C'}\sqrt{1 - j\left(\frac{R'}{\omega L'} + \frac{G'}{\omega C'}\right) - \frac{R'G'}{\omega^2 L'C'}}. \tag{B.10}$$

For small conductor and dielectric losses we can assume that $R'G' \ll \omega^2 L'C'$. Using the first two terms of the Taylor expansion for $\sqrt{1+x} = 1 + x/2 + ...$ we can reduce Eq. (B.10) to

$$\overline{\beta} = \alpha + j\beta \cong j\omega\sqrt{L'C'}\left(1 - \frac{j}{2}\left(\frac{R'}{\omega L'} + \frac{G'}{\omega C'}\right)\right) \quad \text{(B.11)}$$

for low loss transmission lines [127].

The loss coefficient α reads then

$$\alpha \cong \frac{1}{2}\left(R'\sqrt{\frac{C'}{L'}} + G'\sqrt{\frac{L'}{C'}}\right) \quad \text{(B.12)}$$

and the propagation constant β is described with

$$\beta \cong \omega\sqrt{L'C'}. \quad \text{(B.13)}$$

C Electro-optic bandwidth measurement

The bandwidth of an electro-optic modulator is determined using the small signal frequency response of the optical intensity I at the modulation frequency ω_m.

$$S_{21,\mathrm{EO}}(\omega_m) = 10\log_{10}\left(\frac{\delta I(\omega_m)}{\delta I(0)}\right). \tag{C.1}$$

The 3 dB frequency point is the frequency for which the magnitude of the intensity is reduced by half compared to low frequencies. The electro-optic (EO) bandwidth can be determined by measuring the electrical-optical-electrical (EOE) bandwidth of the modulator via a vector network analyzer (VNA) and a calibrated high-speed photo diode [25]. The modulated voltage measured at the 50 Ω input of the VNA $\delta u_{\mathrm{VNA}}(\omega_\mathrm{m})$ is proportional to the current in the photodiode $\delta i_{\mathrm{PD}}(\omega_\mathrm{m})$ and to the intensity of the modulated optical field $\delta I(\omega_\mathrm{m})$. Since the $S_{21,\mathrm{VNA}}$ parameter at the VNA is defined as $20\log_{10}\left(u_{\mathit{VNA}}(\omega_m)/u_{\mathit{VNA}}(0)\right)$, the 3 dB bandwidth of the electro optic response $S_{21,\mathrm{EO}}$ corresponds to the EOE 6 dB bandwidth measured at the VNA.

Appendix

D Field calculations for single-sideband modulation

D.1 Single-sideband modulation without temporal shaping

Frequency shifting by single-sideband modulation can be achieved by utilizing four phase modulators in IQ configuration, see Fig. 4.1 for a schematic. The complex field at the output of a single-sideband modulator can be described as

$$\begin{aligned}\underline{E}_{\text{out}} &= \frac{1}{4}E_0 e^{j\omega_0 t} e^{ja_0 \sin(\Omega t)} - \frac{1}{4}E_0 e^{j\omega_0 t} e^{ja_0 \sin(\Omega t - \pi)} \\ &+ \frac{j}{4}E_0 e^{j\omega_0 t} e^{-jb_0 \cos(\Omega t)} - \frac{j}{4}E_0 e^{j\omega_0 t} e^{-jb_0 \cos(\Omega t - \pi)} \\ &= \frac{1}{4}\underline{E}_{\text{in}}\left(e^{ja_0 \sin(\Omega t)} - e^{-ja_0 \sin(\Omega t)} + je^{-jb_0 \cos(\Omega t)} - je^{jb_0 \cos(\Omega t)}\right).\end{aligned}$$ (D.1)

For an ideal modulator we can assume $a_0 = b_0$, furthermore, the Jacobi-Anger expansion for the expressions of the form

$$e^{jz\cos\theta} = \sum_n j^n J_n(z) e^{jn\theta} \quad \text{and} \quad e^{jz\sin\theta} = \sum_n J_n(z) e^{jn\theta}$$ (D.2)

can be used [118]. $J_n(z)$ is the nth Bessel function of the first kind with $J_n(-z) = (-1)^n J_n(z)$. Using these relations Eq. (D.1) can be written as

$$E_{\text{out}} = \frac{1}{4}E_{\text{in}}\left(\begin{array}{c}\sum_n J_n(a_0)e^{jn\Omega t} - \sum_n (-1)^n J_n(a_0)e^{jn\Omega t} \\ + j\sum_n (-j)^n J_n(a_0)e^{jn\Omega t} - j\sum_n j^n J_n(a_0)e^{jn\Omega t}\end{array}\right) \quad n \in \mathbb{Z}$$ (D.3)

The bracket in Eq. (D.3) gives only non-zero contributions if the relation $1-(-1)^m + j(-j)^m - j(j)^m \neq 0$ holds, which is only the case for $m = 4n+1$. Eq. (D.3) simplifies therefore to

$$E_{\text{out}} = \sum_{m=4n+1} E_0 J_m(a_0) e^{j(\omega_0 + m\Omega)t} \quad n \in \mathbb{Z}.$$ (D.4)

D.2 Single-sideband modulation with additional harmonics

By adding additional harmonics to the drive signal of the single-sideband modulator, spurious side modes can be suppressed. In the following the third harmonic with the modulation amplitude c_0 is added exemplarily to the drive signal,

$$\Phi = a_0 \sin(\Omega t) - c_0 \sin(3\Omega t) \quad \text{and} \quad \Phi = -a_0 \cos(\Omega t) - c_0 \cos(3\Omega t). \tag{D.5}$$

The output field changes now to

$$E_{\text{out}} = \frac{1}{4} E_{\text{in}} \begin{pmatrix} e^{ja_0 \sin(\Omega t)} e^{-jc_0 \sin(3\Omega t)} - e^{-ja_0 \sin(\Omega t)} e^{jc_0 \sin(3\Omega t)} \\ + je^{-ja_0 \cos(\Omega t)} e^{-jc_0 \cos(3\Omega t)} - je^{ja_0 \cos(\Omega t)} e^{jc_0 \cos(3\Omega t)} \end{pmatrix}. \tag{D.6}$$

Using again the Jacobi-Anger expansion the output field can be written as

$$E_{\text{out}} = \frac{1}{4} E_{\text{in}} \begin{pmatrix} \sum_n \sum_k J_n(a_0) e^{jn\Omega t} (-1)^k J_k(c_0) e^{jk3\Omega t} \\ -\sum_n \sum_k (-1)^n J_n(a_0) e^{jn\Omega t} J_k(c_0) e^{jk3\Omega t} \\ +j\sum_n \sum_k (-j)^n J_n(a_0) e^{jn\Omega t} (-j)^k J_k(c_0) e^{jk3\Omega t} \\ -j\sum_n \sum_k j^n J_n(a_0) e^{jn\Omega t} j^k J_k(c_0) e^{jk3\Omega t} \end{pmatrix} \quad n \in \mathbb{Z}. \tag{D.7}$$

If we limit the modulation amplitudes to $a_0, c_0 < 1.84$ we find that only the products

$$J_0(a_0) \sum_k J_k(c_0) \quad \text{and} \quad J_0(c_0) \sum_n J_n(a_0) \tag{D.8}$$

provide significant contributions to the overall sum in Eq. (D.7). Using this relation Eq. (D.7) can be rewritten as

Appendix

$$E_{out} = \frac{1}{4} E_{in} \begin{pmatrix} J_0(c_0)\sum_n J_n(a_0)e^{jn\Omega t} + J_0(a_0)\sum_k (-1)^k J_k(c_0)e^{jk3\Omega t} \\ -J_0(c_0)\sum_n (-1)^n J_n(a_0)e^{jn\Omega t} - J_0(a_0)\sum_k J_k(c_0)e^{jk3\Omega t} \\ +jJ_0(c_0)\sum_n (-j)^n J_n(a_0)e^{jn\Omega t} + jJ_0(a_0)\sum_k (-j)^k J_k(c_0)e^{jk3\Omega t} \\ -jJ_0(c_0)\sum_n j^n J_n(a_0)e^{jn\Omega t} - jJ_0(a_0)\sum_k j^k J_k(c_0)e^{jk3\Omega t} \end{pmatrix}. \quad (D.9)$$

Similar to Eq. (D.3) we find, that the bracket in Eq. gives only non-zero contributions when the relation $1-(-1)^m + j(-j)^m - j(j)^m \neq 0$ holds for $m = 4n+1$ or $1-(-1)^l + j(j)^l - j(-j)^l \neq 0$ for $m = 4k-1$. Eq. (D.9) can therefore be written as

$$E_{out} = E_0 \begin{pmatrix} J_0(c_0) \sum_{m=4n+1} J_m(a_0)e^{j(\omega_0 + m\Omega)t} \\ -J_0(a_0) \sum_{l=4k-1} J_l(c_0)e^{j(\omega_0 + l3\Omega)t} \end{pmatrix}. \quad (D.10)$$

We can now write the significant components of the sum

$$E_{out} = \begin{pmatrix} \ldots + J_0(c_0)J_1(a_0)e^{j(\omega_0+\Omega)t} + J_0(c_0)J_{-3}(a_0)e^{j(\omega_0-3\Omega)t} \\ -J_0(a_0)J_{-1}(c_0)e^{j(\omega_0-3\Omega)t} + J_0(c_0)J_5(a_0)e^{j(\omega_0+5\Omega)t} + \ldots \end{pmatrix}. \quad (D.11)$$

And with a proper choice of a_0 and c_0 the majority of the field at the frequency $(\omega_0 - 3\Omega)$ can be suppressed by fulfilling the equation

$$J_0(c_0)J_{-3}(a_0)e^{j(\omega_0-3\Omega)t} - J_0(a_0)J_{-1}(c_0)e^{j(\omega_0-3\Omega)t} = 0. \quad (D.12)$$

E Bibliography

1. L. Chrostowski and M. Hochberg, *Silicon Photonics Design* (Cambridge University Press, 2015).

2. S. Mittal, "A Survey of Architectural Techniques for Near-Threshold Computing," ACM Journal on Emerging Technologies in Computing Systems **12**, 1–26 (2015).

3. M. Smit, J. van der Tol, and M. Hill, "Moore's law in photonics," Laser Photon. Rev. **6**, 1–13 (2012).

4. M. Smit, X. Leijtens, H. Ambrosius, E. Bente, J. van der Tol, B. Smalbrugge, T. de Vries, E.-J. Geluk, J. Bolk, R. van Veldhoven, L. Augustin, P. Thijs, D. D'Agostino, H. Rabbani, K. Lawniczuk, S. Stopinski, S. Tahvili, A. Corradi, E. Kleijn, D. Dzibrou, M. Felicetti, E. Bitincka, V. Moskalenko, J. Zhao, R. Santos, G. Gilardi, W. Yao, K. Williams, P. Stabile, P. Kuindersma, J. Pello, S. Bhat, Y. Jiao, D. Heiss, G. Roelkens, M. Wale, P. Firth, F. Soares, N. Grote, M. Schell, H. Debregeas, M. Achouche, J.-L. Gentner, A. Bakker, T. Korthorst, D. Gallagher, A. Dabbs, A. Melloni, F. Morichetti, D. Melati, A. Wonfor, R. Penty, R. Broeke, B. Musk, and D. Robbins, "An introduction to InP-based generic integration technology," Semicond. Sci. Technol. **29**, 83001 (2014).

5. C. R. Doerr, "Silicon photonic integration in telecommunications," Frontiers in Physics **3**, (2015).

6. S. Assefa, F. Xia, S. W. Bedell, Y. Zhang, T. Topuria, P. M. Rice, and Y. A. Vlasov, "CMOS-Integrated 40GHz Germanium Waveguide Photodetector for On-Chip Optical Interconnects," in *Optical Fiber Communication Conference* (OSA, 2009), p. OMR4.

7. G. T. Reed, G. Mashanovich, F. Y. Gardes, and D. J. Thomson, "Silicon optical modulators," Nature Photon. **4**, 518–526 (2010).

8. M. Hochberg, N. C. Harris, Ran Ding, Yi Zhang, A. Novack, Zhe Xuan, and T. Baehr-Jones, "Silicon photonics: The next fabless semiconductor industry," IEEE Solid State Circuits Mag. **5**, 48–58 (2013).

9. "Cisco global cloud index: Forecast and methodology, 2015 - 2020," http://www.cisco.com/c/en/us/solutions/collateral/service-provider/global-cloud-index-gci/Cloud_Index_White_Paper.pdf.

10. R.-J. Essiambre, G. Kramer, P. J. Winzer, G. J. Foschini, and B. Goebel, "Capacity limits of optical fiber networks," J. Lightw. Technol. **28**, 662–701 (2010).

11. P. Dong, X. Liu, S. Chandrasekhar, L. L. Buhl, R. Aroca, and Y.-K. Chen, "Monolithic silicon photonic integrated circuits for compact 100+ Gb/s coherent optical receivers and transmitters," IEEE J. Sel. Topics Quantum Electron. **20**, 1–8 (2014).

12. A. Shastri, C. Muzio, M. Webster, G. Jeans, P. Metz, S. Sunder, B. Chattin, B. Dama, and K. Shastri, "Ultra-Low-Power Single-Polarization QAM-16 Generation Without DAC Using a CMOS Photonics Based Segmented Modulator," J. Lightw. Technol. **33**, 1255–1260 (2015).

13. J. Geyer, C. R. Doerr, M. Aydinlik, N. Nadarajah, A. Caballero, C. Rasmussen, and B. Mikkelsen, "Practical implementation of higher order modulation beyond 16-QAM," in *Opt. Fiber Commun. Conf. 2015* (OSA, 2015), p. Th1B.1.

14. T. Baehr-Jones, M. Hochberg, G. Wang, R. Lawson, Y. Liao, P. A. Sullivan, L. Dalton, A. K.-Y. Jen, and A. Scherer, "Optical modulation and detection in slotted silicon waveguides," Opt. Express **13**, 5216 (2005).

15. L. Alloatti, D. Korn, R. Palmer, D. Hillerkuss, J. Li, A. Barklund, R. Dinu, J. Wieland, M. Fournier, J. Fedeli, H. Yu, W. Bogaerts, P. Dumon, R. Baets, C. Koos, W. Freude, and J. Leuthold, "42.7 Gbit/s electro-optic modulator in silicon technology," Opt. Express **19**, 11841 (2011).

16. D. Korn, R. Palmer, H. Yu, P. C. Schindler, L. Alloatti, M. Baier, R. Schmogrow, W. Bogaerts, S. K. Selvaraja, G. Lepage, M. Pantouvaki, J. M. D. Wouters, P. Verheyen, J. Van Campenhout, B. Chen, R. Baets, P. Absil, R. Dinu, C. Koos, W. Freude, and J. Leuthold, "Silicon-organic hybrid (SOH) IQ modulator using the linear electro-optic effect for transmitting 16QAM at 112 Gbit/s," Opt. Express **21**, 13219 (2013).

17. S. Koeber, R. Palmer, M. Lauermann, W. Heni, D. L. Elder, D. Korn, M. Woessner, L. Alloatti, S. Koenig, P. C. Schindler, H. Yu, W. Bogaerts, L. R. Dalton, W. Freude, J. Leuthold, and C. Koos, "Femtojoule electro-optic modulation using a silicon–organic hybrid device," Light Sci. Appl. **4**, e255 (2015).

18. R. Palmer, S. Koeber, D. Elder, M. Woessner, W. Heni, D. Korn, M. Lauermann, W. Bogaerts, L. Dalton, W. Freude, J. Leuthold, and C. Koos, "High-speed, low drive-voltage silicon-organic hybrid modulator based on a binary-chromophore electro-optic material," J. Lightw. Technol. **32**, 2726–2734 (2014).

19. S. Ura, T. Suhara, H. Nishihara, and J. Koyama, "An integrated-optic disk pickup device," J. Lightw. Technol. **4**, 913–918 (1986).

20. A. Rubiyanto, H. Herrmann, R. Ricken, F. Tian, and W. Sohler, "Integrated optical heterodyne interferometer in lithium niboate," J. Nonlinear Opt. Phys. Mater. **10**, 163–168 (2001).

21. L. M. Johnson and C. H. Cox, "Serrodyne optical frequency translation with high sideband suppression," J. Lightw. Technol. **6**, 109–112 (1988).

22. M. Izutsu, S. Shikama, and T. Sueta, "Integrated optical SSB modulator/frequency shifter," IEEE Journal of Quantum Electronics **17**, 2225–2227 (1981).

23. Y. Li, S. Meersman, and R. Baets, "Optical frequency shifter on SOI using thermo-optic serrodyne modulation," in *Group IV Photonics (GFP)* (IEEE, 2010), pp. 75–77.

24. Y. Li, S. Verstuyft, G. Yurtsever, S. Keyvaninia, G. Roelkens, D. Van Thourhout, and R. Baets, "Heterodyne laser Doppler vibrometers integrated on silicon-on-insulator based on serrodyne thermo-optic frequency shifters," Applied Optics **52**, 2145 (2013).

25. R. Palmer, *Silicon Photonic Modulators for Low-Power Applications* (KIT Scientific Publishing, 2015).

26. R. W. Boyd, *Nonlinear Optics*, 3rd ed (Academic Press, 2008).

27. M. Seimetz, *High-Order Modulation for Optical Fiber Transmission*, Springer Series in Optical Sciences No. 143 (Springer, 2009).

28. G.-W. Lu, M. Sköld, P. Johannisson, J. Zhao, M. Sjödin, H. Sunnerud, M. Westlund, A. Ellis, and P. A. Andrekson, "40-Gbaud 16-QAM transmitter using tandem IQ modulators with binary driving electronic signals," Optics Express **18**, 23062 (2010).

29. R. S. Jacobsen, K. N. Andersen, P. I. Borel, J. Fage-Pedersen, L. H. Frandsen, O. Hansen, M. Kristensen, A. V. Lavrinenko, G. Moulin, H. Ou, C. Peucheret, B. Zsigri, and A. Bjarklev, "Strained silicon as a new electro-optic material," Nature **441**, 199–202 (2006).

30. P. Damas, X. Le Roux, D. Le Bourdais, E. Cassan, D. Marris-Morini, N. Izard, T. Maroutian, P. Lecoeur, and L. Vivien, "Wavelength dependence of Pockels effect in strained silicon waveguides," Optics Express **22**, 22095 (2014).

31. S. Sharif Azadeh, F. Merget, M. P. Nezhad, and J. Witzens, "On the measurement of the Pockels effect in strained silicon," Optics Letters **40**, 1877 (2015).

32. C. Koos, J. Brosi, M. Waldow, W. Freude, and J. Leuthold, "Silicon-on-insulator modulators for next-generation 100 Gbit/s-ethernet," in *33rd European Conference and Exhibition of Optical Communication (ECOC)* (VDE, 2007).

33. J. Leuthold, C. Koos, W. Freude, L. Alloatti, R. Palmer, D. Korn, J. Pfeifle, M. Lauermann, R. Dinu, S. Wehrli, M. Jazbinsek, P. Gunter, M. Waldow, T. Wahlbrink, J. Bolten, H. Kurz, M. Fournier, J.-M. Fedeli, H. Yu, and W. Bogaerts, "Silicon-organic hybrid electro-optical devices," IEEE J. Sel. Topics Quantum Electron. **19**, 114–126 (2013).

34. J. Liu, *Photonic Devices* (Cambridge University Press, 2005).

35. G. P. Agrawal, *Lightwave Technology: Components and Devices* (John Wiley, 2004).

36. A. K. Bain, T. J. Jackson, Y. Koutsonas, M. Cryan, S. Yu, M. Hill, R. Varrazza, J. Rorison, and M. J. Lancaster, "Optical Properties of Barium Strontium Titanate (BST) Ferroelectric Thin Films," Ferroelectrics Letters Section **34**, 149–154 (2007).

37. C. B. Parker, J.-P. Maria, and A. I. Kingon, "Temperature and thickness dependent permittivity of $(Ba,Sr)TiO_3$ thin films," Appl. Phys. Lett. **81**, 340 (2002).

38. J. Witzens, T. Baehr-Jones, and M. Hochberg, "Design of transmission line driven slot waveguide Mach-Zehnder interferometers and application to analog optical links," Opt. Express **18**, 16902 (2010).

39. L. R. Dalton, P. A. Sullivan, and D. H. Bale, "Electric field poled organic electro-optic materials: state of the art and future prospects," Chem. Rev. **110**, 25–55 (2010).

40. J. J. Wolff and R. Wortmann, "Organic Materials for Second-Order Non-Linear Optics," in *Advances in Physical Organic Chemistry* (Elsevier, 1999), Vol. 32, pp. 121–217.

41. G. Gupta, W. H. Steier, Y. Liao, J. Luo, L. R. Dalton, and A. K.-Y. Jen, "Modeling Photobleaching of Optical Chromophores: Light-Intensity Effects in Precise Trimming of Integrated Polymer Devices," J. Phys. Chem. C **112**, 8051–8060 (2008).

42. Y. Ren, M. Szablewski, and G. H. Cross, "Waveguide photodegradation of nonlinear optical organic chromophores in polymeric films," Applied Optics **39**, 2499 (2000).

43. P. Kaatz, P. Prêtre, U. Meier, U. Stalder, C. Bosshard, P. Günter, B. Zysset, M. Stähelin, M. Ahlheim, and F. Lehr, "Relaxation Processes in Nonlinear Optical Polyimide Side-Chain Polymers," Macromolecules **29**, 1666–1678 (1996).

44. G. Williams and D. C. Watts, "Non-symmetrical dielectric relaxation behaviour arising from a simple empirical decay function," Trans. Faraday Soc. **66**, 80 (1970).

45. M. Haller, J. Luo, H. Li, T.-D. Kim, Y. Liao, B. H. Robinson, L. R. Dalton, and A. K.-Y. Jen, "A Novel Lattice-Hardening Process To Achieve Highly Efficient and Thermally Stable Nonlinear Optical Polymers," Macromolecules **37**, 688–690 (2004).

46. J. Luo, S. Huang, Z. Shi, B. M. Polishak, X.-H. Zhou, and A. K. Jen, "Tailored organic electro-optic materials and their hybrid systems for device applications," Chem. Mat. **23**, 544–553 (2011).

47. Y. Enami, J. Luo, and A. K.-Y. Jen, "Short hybrid polymer/sol-gel silica waveguide switches with high in-device electro-optic coefficient based on photostable chromophore," AIP Adv. **1**, 42137-42137–7 (2011).

48. Z. Shi, J. Luo, S. Huang, B. M. Polishak, X.-H. Zhou, S. Liff, T. R. Younkin, B. A. Block, and A. K.-Y. Jen, "Achieving excellent electro-optic activity and thermal stability in poled polymers through an expeditious crosslinking process," J. Mater. Chem. **22**, 951 (2012).

49. R. Soref and B. Bennett, "Electrooptical effects in silicon," IEEE J. Quantum Electron. **23**, 123–129 (1987).

50. M. R. Watts, W. A. Zortman, D. C. Trotter, R. W. Young, and A. L. Lentine, "Low-voltage, compact, depletion-mode, silicon Mach-Zehnder modulator," IEEE J. Sel. Topics Quantum Electron. **16**, 159–164 (2010).

51. P. Dong, C. Xie, L. L. Buhl, Y.-K. Chen, J. H. Sinsky, and G. Raybon, "Silicon in-phase/quadrature modulator with on-chip optical equalizer," J. Lightw. Technol. **33**, 1191–1196 (2015).

52. W. M. Green, M. J. Rooks, L. Sekaric, and Y. A. Vlasov, "Ultra-compact, low RF power, 10 Gb/s silicon Mach-Zehnder modulator," Opt. Express **15**, 17106 (2007).

53. T. Baba, S. Akiyama, M. Imai, T. Akagawa, M. Takahashi, N. Hirayama, H. Takahashi, Y. Noguchi, H. Okayama, T. Horikawa, and T. Usuki, "25-Gbps operation of silicon p-i-n Mach-Zehnder optical modulator with 100-µm-long phase shifter," in *CLEO: Science and Innovations* (OSA, 2012), p. CF2L.3.

54. J. Fujikata, J. Ushida, T. Nakamura, Y. Ming-Bin, Z. ShiYang, D. Liang, P. L. Guo-Qiang, and D.-L. Kwong, "25 GHz operation of silicon optical modulator with projection MOS structure," in *Optical Fiber Communication Conference* (OSA, 2010), p. OMI.3.

55. B. Milivojevic, C. Raabe, A. Shastri, M. Webster, P. Metz, S. Sunder, B. Chattin, S. Wiese, B. Dama, and K. Shastri, "112Gb/s DP-QPSK transmission over 2427km SSMF using small-size silicon photonic IQ modulator and low-power CMOS driver," in *Optical Fiber Communication Conference, 2013* (OSA, 2013), p. OTh1D.1.

56. C. Koos, J. Leuthold, W. Freude, M. Kohl, L. Dalton, W. Bogaerts, A. Giesecke, M. Lauermann, A. Melikyan, S. Koeber, S. Wolf, C. Weimann, S. Muehlbrandt, K. Koehnle, J. Pfeifle, W. Hartmann, Y. Kutuvantavida, S. Ummethala, R. Palmer, D. Korn, L. Alloatti, P. Schindler, D. Elder, T. Wahlbrink, and J. Bolten, "Silicon-organic hybrid (SOH) and plasmonic-organic hybrid (POH) integration," J. Lightw. Technol. **34**, 256–268 (2016).

57. T. Baehr-Jones, B. Penkov, J. Huang, P. Sullivan, J. Davies, J. Takayesu, J. Luo, T.-D. Kim, L. Dalton, A. Jen, M. Hochberg, and A. Scherer, "Nonlinear polymer-clad silicon slot waveguide modulator with a half wave voltage of 0.25 V," Appl. Phys. Lett. **92**, 163303 (2008).

58. J. H. Wülbern, S. Prorok, J. Hampe, A. Petrov, M. Eich, J. Luo, A. K.-Y. Jen, M. Jenett, and A. Jacob, "40 GHz electro-optic modulation in hybrid silicon–organic slotted photonic crystal waveguides," Opt. Lett. **35**, 2753 (2010).

59. K. Goi, A. Oka, H. Kusaka, Y. Terada, K. Ogawa, T.-Y. Liow, X. Tu, G.-Q. Lo, and D.-L. Kwong, "Low-loss high-speed silicon IQ modulator for QPSK/DQPSK in C and L bands," Opt. Express **22**, 10703 (2014).

60. X. Zhang, A. Hosseini, X. Lin, H. Subbaraman, and R. T. Chen, "Polymer-based hybrid-integrated photonic devices for silicon on-chip modulation and board-level optical interconnects," IEEE J. Sel. Topics Quantum Electron. **19**, 196–210 (2013).

61. B. Chmielak, M. Waldow, C. Matheisen, C. Ripperda, J. Bolten, T. Wahlbrink, M. Nagel, F. Merget, and H. Kurz, "Pockels effect based fully integrated, strained silicon electro-optic modulator," Opt. Express **19**, 17212 (2011).

62. J.-M. Brosi, C. Koos, L. C. Andreani, M. Waldow, J. Leuthold, and W. Freude, "High-speed low-voltage electro-optic modulator with a polymer-infiltrated silicon photonic crystal waveguide," Opt. Express **16**, 4177 (2008).

63. X. Zhang, H. Subbaraman, A. Hosseini, and R. T. Chen, "Highly efficient mode converter for coupling light into wide slot photonic crystal waveguide," Opt. Express **22**, 20678 (2014).

64. X. Zhang, A. Hosseini, S. Chakravarty, J. Luo, A. K.-Y. Jen, and R. T. Chen, "Wide optical spectrum range, subvolt, compact modulator based on an electro-optic polymer refilled silicon slot photonic crystal waveguide," Opt. Lett. **38**, 4931 (2013).

65. R. Palmer, S. Koeber, W. Heni, D. L. Elder, D. Korn, H. Yu, L. Alloatti, S. Koenig, P. C. Schindler, W. Bogaerts, M. Pantouvaki, G. Lepage, P. Verheyen, J. Van Campenhout, P. Absil, R. Baets, L. R. Dalton, W. Freude, J. Leuthold, and C. Koos, "High-speed silicon-organic hybrid (SOH) modulator with 1.6 fJ/bit and 180 pm/V in-device nonlinearity," in *39th European Conference and Exhibition on Optical Communication (ECOC 2013)* (IET, 2013), p. We.3.B.3.

66. Y. V. Pereverzev, K. N. Gunnerson, O. V. Prezhdo, P. A. Sullivan, Y. Liao, B. C. Olbricht, A. J. P. Akelaitis, A. K.-Y. Jen, and L. R. Dalton, "Guest–host cooperativity in organic materials greatly enhances the nonlinear optical response," J. Phys. Chem. C **112**, 4355–4363 (2008).

67. M. Lauermann, R. Palmer, S. Koeber, P. C. Schindler, D. Korn, T. Wahlbrink, J. Bolten, M. Waldow, D. L. Elder, L. R. Dalton, J. Leuthold, W. Freude, and C. G. Koos, "16QAM silicon-organic hybrid (SOH) modulator operating with 0.6 V_{pp} and 19 fJ/bit at 112 Gbit/s," in *CLEO: Science and Innovations* (OSA, 2014), p. SM2G.6.

68. L. Alloatti, M. Lauermann, C. Sürgers, C. Koos, W. Freude, and J. Leuthold, "Optical absorption in silicon layers in the presence of charge inversion/accumulation or ion implantation," Appl. Phys. Lett. **103**, 51104 (2013).

69. Z. Shi, W. Liang, J. Luo, S. Huang, B. M. Polishak, X. Li, T. R. Younkin, B. A. Block, and A. K.-Y. Jen, "Tuning the kinetics and energetics of Diels−Alder cycloaddition reactions to improve poling efficiency and thermal stability of high-temperature cross-linked electro-optic polymers," Chem. Mat. **22**, 5601–5608 (2010).

70. R. Schmogrow, D. Hillerkuss, M. Dreschmann, M. Huebner, M. Winter, J. Meyer, B. Nebendahl, C. Koos, J. Becker, W. Freude, and J. Leuthold, "Real-time software-defined multiformat transmitter generating 64QAM at 28 GBd," IEEE Photon. Technol. Lett. **22**, 1601–1603 (2010).

71. R. Palmer, L. Alloatti, D. Korn, W. Heni, P. C. Schindler, J. Bolten, M. Karl, M. Waldow, T. Wahlbrink, W. Freude, C. Koos, and J. Leuthold, "Low-loss silicon strip-to-slot mode converters," IEEE Photon. J. **5**, 2200409–2200409 (2013).

72. F. Derr, "Coherent optical QPSK intradyne system: concept and digital receiver realization," J. Lightw. Technol. **10**, 1290–1296 (1992).

73. R. Ding, T. Baehr-Jones, W.-J. Kim, B. Boyko, R. Bojko, A. Spott, A. Pomerene, C. Hill, W. Reinhardt, and M. Hochberg, "Low-loss asymmetric strip-loaded slot waveguides in silicon-on-insulator," Appl. Phys. Lett. **98**, 233303 (2011).

74. D. Vermeulen, S. Selvaraja, P. Verheyen, G. Lepage, W. Bogaerts, P. Absil, D. Van Thourhout, and G. Roelkens, "High-efficiency fiber-to-chip grating couplers realized using an advanced CMOS-compatible silicon-on-insulator platform," Opt. Express **18**, 18278 (2010).

75. N. Lindenmann, S. Dottermusch, T. Hoose, M. R. Billah, S. Koeber, W. Freude, and C. Koos, "Connecting silicon photonic circuits to multi-core fibers by photonic wire bonding," in *Optical Interconnects Conference, 2014* (IEEE, 2014), pp. 131–132.

76. N. Lindenmann, S. Dottermusch, M. L. Goedecke, T. Hoose, M. R. Billah, T. Onanuga, A. Hofmann, W. Freude, and C. Koos, "Connecting silicon photonic circuits to multi-core fibers by photonic wire bonding," J. Lightw. Technol. accepted for publication (2014).

77. R. Schmogrow, B. Nebendahl, M. Winter, A. Josten, D. Hillerkuss, S. Koenig, J. Meyer, M. Dreschmann, M. Huebner, C. Koos, J. Becker, W. Freude, and J. Leuthold, "Error vector magnitude as a performance measure for advanced modulation formats," IEEE Photon. Technol. Lett. **24**, 61–63 (2012).

78. J. S. Orcutt, B. Moss, C. Sun, J. Leu, M. Georgas, J. Shainline, E. Zgraggen, H. Li, J. Sun, M. Weaver, S. Urošević, M. Popović, R. J. Ram, and V. Stojanović, "Open foundry platform for high-performance electronic-photonic integration," Opt. Express **20**, 12222 (2012).

79. P. J. Winzer, "High-spectral-efficiency optical modulation formats," J. Lightw. Technol. **30**, 3824–3835 (2012).

80. L. Liao, A. Liu, D. Rubin, J. Basak, Y. Chetrit, H. Nguyen, R. Cohen, N. Izhaky, and M. Paniccia, "40 Gbit/s silicon optical modulator for high-speed applications," Electron. Lett. **43**, 1196–1197 (2007).

81. P. Dong, C. Xie, L. L. Buhl, Y.-K. Chen, J. H. Sinsky, and G. Raybon, "Silicon in-phase/quadrature modulator with on-chip optical equalizer," in *40th European Conference on Optical Communication (ECOC 2014)* (2014), p. We.1.4.5.

82. V. R. Almeida, Q. Xu, C. A. Barrios, and M. Lipson, "Guiding and confining light in void nanostructure," Opt. Lett. **29**, 1209 (2004).

83. R. Palmer, L. Alloatti, D. Korn, P. C. Schindler, R. Schmogrow, W. Heni, S. Koenig, J. Bolten, T. Wahlbrink, M. Waldow, H. Yu, W. Bogaerts, P. Verheyen, G. Lepage, M. Pantouvaki, J. Van Campenhout, P. Absil, R. Dinu, W. Freude, C. Koos, and J. Leuthold, "Silicon-organic hybrid MZI modulator generating OOK, BPSK and 8-ASK signals for up to 84 Gbit/s," IEEE Photon. J. **5**, 6600907 (2013).

84. M. Lauermann, P. C. Schindler, S. Wolf, R. Palmer, S. Koeber, D. Korn, L. Alloatti, T. Wahlbrink, J. Bolten, M. Waldow, M. Koenigsmann, M. Kohler, D. Malsam, D. L. Elder, P. V. Johnston, N. Phillips-Sylvain, P. A. Sullivan, L. R. Dalton, J. Leuthold, W. Freude, and C. Koos, "40 GBd 16QAM modulation at 160 Gbit/s in a silicon-organic hybrid (SOH) modulator," in *40th European Conference on Optical Communication (ECOC 2014)* (2014), p. We.3.1.3.

85. R. Palmer, L. Alloatti, D. Korn, P. C. Schindler, M. Baier, J. Bolten, T. Wahlbrink, M. Waldow, R. Dinu, W. Freude, C. Koos, and J. Leuthold, "Low power Mach-Zehnder modulator in silicon-organic hybrid technology," IEEE Photon. Technol. Lett. **25**, 1226–1229 (2013).

86. L. Alloatti, R. Palmer, S. Diebold, K. P. Pahl, B. Chen, R. Dinu, M. Fournier, J.-M. Fedeli, T. Zwick, W. Freude, C. Koos, and J. Leuthold, "100 GHz silicon–organic hybrid modulator," Light Sci. Appl. **3**, e173 (2014).

87. N. Lindenmann, G. Balthasar, D. Hillerkuss, R. Schmogrow, M. Jordan, J. Leuthold, W. Freude, and C. Koos, "Photonic wire bonding: a novel concept for chip-scale interconnects," Opt. Express **20**, 17667 (2012).

88. F. Chang, K. Onohara, and T. Mizuochi, "Forward error correction for 100 G transport networks," IEEE Commun. Mag. **48**, 48–55 (2010).

89. D. Chang, F. Yu, Z. Xiao, Y. Li, N. Stojanovic, C. Xie, X. Shi, X. Xu, and Q. Xiong, "FPGA verification of a single QC-LDPC code for 100 Gb/s optical systems without error floor down to BER of 10^{-15}," in *Optical Fiber Communication Conference (OFC)* (OSA, 2011), p. OTuN2.

90. R. G. Hunsperger, *Integrated Optics: Theory and Technology*, 6th ed, Advanced Texts in Physics (Springer, 2009).

91. R. Mäusl and J. Göbel, *Analoge und digitale Modulationsverfahren: Basisband und Trägermodulation* (Hüthig, 2002).

92. R. V. L. Hartley, "Modulation system," U.S. patent 1,666,206 (April 17, 1928).

93. R. Cumming, "The Serrodyne Frequency Translator," Proceedings of the IRE **45**, 175–186 (1957).

94. S. Shimotsu, S. Oikawa, T. Saitou, N. Mitsugi, K. Kubodera, T. Kawanishi, and M. Izutsu, "Single side-band modulation performance of a $LiNbO_3$ integrated modulator consisting of four-phase modulator waveguides," IEEE Photonics Technology Letters **13**, 364–366 (2001).

95. I. Y. Poberezhskiy, B. Bortnik, J. Chou, B. Jalali, and H. R. Fetterman, "Serrodyne frequency translation of continuous optical signals using ultrawide-band electrical sawtooth waveforms," IEEE J. Quantum Electron. **41**, 1533–1539 (2005).

96. Y. Li, "Miniaturized laser doppler vibrometer integrated on a silicon photonics platform," Ghent University (2013).

97. N. C. Harris, Y. Ma, J. Mower, T. Baehr-Jones, D. Englund, M. Hochberg, and C. Galland, "Efficient, compact and low loss thermo-optic phase shifter in silicon," Opt. Express **22**, 10487 (2014).

98. M. R. Watts, J. Sun, C. DeRose, D. C. Trotter, R. W. Young, and G. N. Nielson, "Adiabatic thermo-optic Mach–Zehnder switch," Optics Letters **38**, 733 (2013).

99. M. P. Kothiyal and C. Delisle, "Optical frequency shifter for heterodyne interferometry using counterrotating wave plates," Opt. Lett. **9**, 319 (1984).

100. M. Bauer, F. Ritter, and G. Siegmund, "High-precision laser vibrometers based on digital Doppler signal processing," in E. P. Tomasini, ed. (2002), pp. 50–61.

101. Y. Ma, Q. Yang, Y. Tang, S. Chen, and W. Shieh, "1-Tb/s single-channel coherent optical OFDM transmission over 600-km SSMF fiber with subwavelength bandwidth access," Optics Express **17**, 9421 (2009).

102. R. Watts, S. G. Murdoch, and L. Barry, "Spectral linewidth reduction of single-mode and mode-locked lasers using a feed-forward heterodyne detection scheme," in *CLEO: 2014* (OSA, 2014), p. STh3O.8.

103. Y. Park and K. Cho, "Heterodyne interferometer scheme using a double pass in an acousto-optic modulator," Opt. Lett. **36**, 331 (2011).

104. G. Cocorullo and I. Rendina, "Thermo-optical modulation at 1.5µm in silicon etalon," Electron. Lett. **28**, 83–85 (1992).

105. J. Pfeifle, L. Alloatti, W. Freude, J. Leuthold, and C. Koos, "Silicon-organic hybrid phase shifter based on a slot waveguide with a liquid-crystal cladding," Opt. Express **20**, 15359 (2012).

106. Y. Xing, T. Ako, J. P. George, D. Korn, H. Yu, P. Verheyen, M. Pantouvaki, G. Lepage, P. Absil, A. Ruocco, C. Koos, J. Leuthold, K. Neyts, J. Beeckman, and W. Bogaerts, "Digitally controlled phase shifter using an SOI slot waveguide with liquid crystal infiltration," IEEE Photon. Technol. Lett. **27**, 1269–1272 (2015).

107. D. Korn, M. Lauermann, S. Koeber, P. Appel, L. Alloatti, R. Palmer, P. Dumon, W. Freude, J. Leuthold, and C. Koos, "Lasing in silicon–organic hybrid waveguides," Nature Communications **7**, 10864 (2016).

108. F. Qiu, H. Sato, A. M. Spring, D. Maeda, M. Ozawa, K. Odoi, I. Aoki, A. Otomo, and S. Yokoyama, "Ultra-thin silicon/electro-optic polymer hybrid waveguide modulators," Applied Physics Letters **107**, 123302 (2015).

109. S. Inoue and A. Otomo, "Electro-optic polymer/silicon hybrid slow light modulator based on one-dimensional photonic crystal waveguides," Applied Physics Letters **103**, 171101 (2013).

110. M. Gould, T. Baehr-Jones, R. Ding, S. Huang, J. Luo, A. K.-Y. Jen, J.-M. Fedeli, M. Fournier, and M. Hochberg, "Silicon-polymer hybrid slot waveguide ring-resonator modulator," Opt. Express **19**, 3952 (2011).

111. M. Lauermann, C. Weimann, A. Knopf, D. L. Elder, W. Heni, R. Palmer, D. Korn, P. Schindler, S. Koeber, L. Alloatti, H. Yu, W. Bogaerts, L. R. Dalton, C. Rembe, J. Leuthold, W. Freude, and C. Koos, "Integrated Silicon-Organic Hybrid (SOH) Frequency Shifter," in (Optical Fiber Communication Conference, 2014), p. Tu2A.1.

112. X. Zhang, C. Chung, A. Hosseini, H. Subbaraman, J. Luo, A. Jen, R. Neilson, C. Lee, and R. T. Chen, "High performance optical modulator based on electro-optic polymer filled silicon slot photonic crystal waveguide," J. Lightw. Technol. 1–1 (2015).

113. W. Hartmann, M. Lauermann, S. Wolf, H. Zwickel, Y. Kutuvantavida, J. Luo, A. K.-Y. Jen, W. Freude, and C. Koos, "100 Gbit/s OOK using a silicon-organic hybrid (SOH) modulator," in *European Conference on Optical Communication (ECOC), 2015* (IEEE, 2015), p. PDP.1.4.

114. D. L. Elder, S. J. Benight, J. Song, B. H. Robinson, and L. R. Dalton, "Matrix-assisted poling of monolithic bridge-disubstituted organic NLO chromophores," Chem. Mat. **26**, 872–874 (2014).

115. D. Benedikovic, C. Alonso-Ramos, P. Cheben, J. H. Schmid, S. Wang, D.-X. Xu, J. Lapointe, S. Janz, R. Halir, A. Ortega-Moñux, J. G. Wangüemert-Pérez, I. Molina-Fernández, J.-M. Fédéli, L. Vivien, and M. Dado, "High-directionality fiber-chip grating coupler with interleaved trenches and subwavelength index-matching structure," Optics Letters **40**, 4190 (2015).

116. M. Lauermann, S. Wolf, P. C. Schindler, R. Palmer, S. Koeber, D. Korn, L. Alloatti, T. Wahlbrink, J. Bolten, M. Waldow, M. Koenigsmann, M. Kohler, D. Malsam, D. L. Elder, P. V. Johnston, N. Phillips-Sylvain, P. A. Sullivan, L. R. Dalton, J. Leuthold, W. Freude, and C. Koos, "40 GBd 16QAM signaling at 160 Gb/s in a silicon-organic hybrid modulator," J. Lightw. Technol. **33**, 1210–1216 (2015).

117. J. P. Salvestrini, L. Guilbert, M. Fontana, M. Abarkan, and S. Gille, "Analysis and control of the DC drift in LiNbO3-based Mach-Zehnder modulators," J. Lightw. Technol. **29**, 1522–1534 (2011).

118. I. S. Gradshteyn and I. M. Ryzhik, *Tables of Series, Products, and Integrals*, 2nd ed. (Harri Deutsch, 1981).

119. S. Huang, J. Luo, Z. Jin, X.-H. Zhou, Z. Shi, and A. K.-Y. Jen, "Enhanced temporal stability of a highly efficient guest–host electro-optic polymer through a barrier layer assisted poling process," J. Mater. Chem. **22**, 20353 (2012).

120. H. Xu, X. Li, X. Xiao, P. Zhou, Z. Li, J. Yu, and Y. Yu, "High speed silicon modulator with band equalization," Opt. Lett. **39**, 4839 (2014).

121. Y. Fang, L. Liu, C. Y. Wong, S. Zhang, T. Wang, G. N. Liu, and X. Xu, "Silicon IQ modulator based 480km 80x453.2Gb/s PDM-eOFDM transmission on 50GHz grid with SSMF and EDFA-only link," in *Opt. Fiber Commun. Conf. 2015* (OSA, 2015), p. M3G.5.

122. M. Lauermann, R. Palmer, S. Koeber, P. C. Schindler, D. Korn, T. Wahlbrink, J. Bolten, M. Waldow, D. L. Elder, L. R. Dalton, J. Leuthold, W. Freude, and C. Koos, "Low-power silicon-organic hybrid (SOH) modulators for advanced modulation formats," Opt. Express **22**, 29927 (2014).

123. S. Wolf, M. Lauermann, P. Schindler, G. Ronniger, K. Geistert, R. Palmer, S. Kober, W. Bogaerts, J. Leuthold, W. Freude, and C. Koos, "DAC-less amplifier-less generation and transmission of QAM signals using sub-volt silicon-organic hybrid modulators," J. Lightw. Technol. **33**, 1425–1432 (2015).

124. M. Lauermann, S. Wolf, R. Palmer, A. Bielik, L. Altenhain, J. Lutz, R. Schmid, T. Wahlbrink, J. Bolten, A. L. Giesecke, W. Freude, and C. Koos, "64 GBd operation of a silicon-organic hybrid modulator at elevated temperature," in *Optical Fiber Communication Conference, 2015* (OSA, 2015), p. Tu2A.5.

125. R. Ding, T. Baehr-Jones, W.-J. Kim, X. Xiong, R. Bojko, J.-M. Fedeli, M. Fournier, and M. Hochberg, "Low-loss strip-loaded slot waveguides in Silicon-on-Insulator," Opt. Express **18**, 25061 (2010).

126. G. K. Grau and W. Freude, *Optische Nachrichtentechnik: Eine Einführung*, 3rd ed. (Springer-Verlag, 1991).

127. D. M. Pozar, *Microwave Engineering*, 4th ed (Wiley, 2012).

Appendix

F Glossary

F.1 List of abbreviations

16QAM	16-state quadrature amplitude modulation
4ASK	4-state amplitude shift keying
ASE	Amplified spontaneous emission
ASK	Amplitude shift keying
AOM	Acousto-optic modulator
AWGN	Additive white Gaussian noise
AWG	Arbitrary waveform generator
BCOG	binary-chromophore organic glass
BER	Bit error ratio
BOX	Buried oxide
BPSK	Binary phase shift keying
CE	Conversion efficiency
CS	Carrier suppression
CMOS	Complementary metal-oxide-semiconductor
DAC	Digital-to-analog converter
DUV	deep ultraviolet
ECL	External cavity laser
EDFA	Erbium-doped fiber amplifier
EO	Electro-optic
EOE	Electro-optic-electric
EVM	Error vector magnitude

FEC	Forward error correction
FPGA	Field-programmable gate arrays
GC	Grating coupler
GSG	Ground – signal – ground
HSQ	Hydrogen silsesquioxane
IME	Institute of Microelectronics A*Star
InP	Indium phosphide
IQ	in-phase / quadrature phase
$LiNbO_3$	Lithium Niobate
LO	Local oscillator
MMI	Multimode interference coupler
MOS	Metal-oxide-semiconductor
MZM	Mach-Zehnder modulator
NRZ	Non-return to zero
OMA	Optical modulation analyzer
OOK	On-off keying
OSNR	Optical signal to noise ratio
PDK	Process design kit
PIC	Photonic integrated circuit
PM	Phase modulator
PRBS	Pseudorandom binary sequence
PSK	Phase shift keying
QAM	Quadrature amplitude modulation
QPSK	Quadrature phase shift keying
RF	Radio frequency
Si	Silicon

SMSR	Side-mode suppression ratio
SNR	Signal to noise ratio
SOH	Silicon organic hybrid
SOI	Silicon on insulator
SSB	Single-sideband
VNA	Vector network analyzer

F.2 List of mathematical symbols

Greek symbols

α	Loss parameter
α_{ij}	Linear polarizability
β	Propagation constant
β_{ijk}	Second-order polarizability
γ	Weighting factor for the field interaction
γ_{ijkl}	Third order polarizability
Γ	Field interaction factor
Γ_R	Reflection coefficient
ε_0	Vacuum permittivity
ε_r	Relative dielectric permittivity
$\underline{\varepsilon}_r$	Relative dielectric permittivity tensor
$\Delta\varepsilon_r$	Change of the relative permittivity
κ	Complex part of the refractive index
λ	Wavelength
μ_0	Vacuum permeability
φ	Phase of a signal

Φ	Phase modulation
ψ	Relaxation process described by a stretched exponential function
$\underline{\chi}^{(n)}$	Response function of order n
$\underline{\tilde{\chi}}^{(n)}$	Electric susceptibility of order n
ω	Angular frequency
ω_0	Angular frequency of the optical carrier
Ω	Angular modulation frequency

Latin symbols

a_0	Modulation amplitude
A	Alphabet of symbols
A_{EO}	Cross-sectional area filled with electro-optic material
A_{slot}	Cross-sectional area of the slot waveguide, $h_{\mathrm{Si}} \times w_{\mathrm{slot}}$
b_0	Modulation amplitude
b_k	Complex symbol for data transmission
b_k^{i}	Real part of a symbol
b_k^{q}	Complex part of a symbol
c_0	Modulation amplitude
c	Speed of light
C	Capacitance
$\vec{D}$	Electric displacement field vector
$\vec{E}$	Electrical field vector
$\vec{\mathcal{E}}$	Transversal electric mode field
g	Lorentz-Onsager field factor
h_{slab}	Height of the doped silicon slabs
h_{Si}	Height of the silicon device layer

$i(t)$	Time dependent current
$\vec{\mathcal{H}}$	Transversal magnetic mode field
I	Optical intensity
J_n	nth Bessel function of the first kind
L	Length
M	Number of elements in an alphabet of symbols
$\bar{n}$	Complex refractive index
n	Real part of the refractive index
N	Chromophore density
$\vec{P}$	Electric polarization
r_s	Symbol rate
r_{ijk}	Electro-optic coefficient, the indices i,j,k denote the Cartesian components
$\vec{r}$	Radius
R	Resistance
s_a	Complex analytical baseband signal
s_{BB}	Real valued baseband signal
s_{SSB}	Complex single sideband signal
t	Time
T	Period of a signal
T_f	Fall time
T_g	Glass transition temperature
T_S	Symbol duration
$u(t)$	Time dependent voltage
U_D	Drive amplitude
U_π	π-voltage

v_g	Group velocity
w_{rail}	Width of the silicon rails
w_{slot}	Width of the silicon slot
Z	Characteristic impedance
Z_0	Impedance of free space

Danksagungen

Diese Dissertation entstand während meiner Zeit am Institut für Photonik und Quantenelektronik (IPQ) am Karlsruher Institut für Technologie. Die Arbeit war eingebettet in die Forschungsprojekte PhoxTroT und BigPipes der Europäischen Union und wurde gefördert von der Helmholtz International Research School for Teratronics.

An dieser Stelle darf ich mich bei allen Personen bedanken, die mich in den letzten Jahren unterstützt und diese Arbeit erst ermöglicht haben.

Als erstes möchte ich mich herzlich bei meinem Doktorvater Prof. Christian Koos bedanken. Sein Ideenreichtum und Tatendrang waren immer motivierend und die offenen Diskussionen über Siliziumphotonik haben wesentlich dazu beigetragen die Arbeit auch im größeren Zusammenhang zu sehen. Vielen Dank für konstante Unterstützung über all die Jahre, ohne die die vielfältigen Experimente und Arbeiten nicht möglich gewesen wären. Vielen Dank auch an Prof. Wolfgang Freude für die fruchtbaren Diskussionen und die gründlichen Korrekturen der Ausarbeitungen. Es war immer sehr lehrreich auftretende Probleme der täglichen Arbeit detailliert und von verschiedenen Blickwinkeln gemeinsam anzugehen und zu lösen. Prof. Jürg Leuthold möchte ich für seine Inspiration und Motivation danken, die mich für die Photonik begeistert hat und schließlich auf den Weg der integrierten Optik gebracht hat. Die Zusammenarbeit mit Prof. Christian Rembe hat mir die Messtechnik nähergebracht und wesentlich dazu beigetragen, dass diese Arbeit mit dem Frequenzschieber um einen bedeutenden Aspekt bereichert wurde, vielen Dank dafür.

Die Arbeit im Rahmen des Forschungsprojekt PhoxTroT bot die Gelegenheit mit vielen großartigen Kollegen aus ganz Europa zusammenzuarbeiten. Besonders bedanken möchte ich mich hier bei Thorsten Wahlbrink, Anna-Lena Giesecke und Jens Bolten von der AMO GmbH sowie Sander Dorrestein, Rutger Smink und Jeroen Duis von TE Connectivity für die gute Zusammenarbeit. Es war extrem spannend und es hat mir viel Freude bereitet gemeinsam die SOH Modulatoren aus der Universität in Richtung Anwendung zu bringen.

Danksagungen

Weiterhin möchte ich mich bei Prof. Jeremy Witzens und Florian Merget von der RWTH Aachen für die Organisation der Chipfertigung bei IME im Rahmen des BigPipes Projektes bedanken. Vielen Dank auch für die Unterstützung beim Erstellen der Maskenlayouts und bei der Definition der Prozessdetails.

Vielen Dank an Prof. Larry Dalton von der University of Washington und sein Team. Die elektrooptischen Materialien aus seinem Labor setzen Maßstäbe im Hinblick auf ihre Effizienz und haben zahlreiche Experimente ermöglicht. Weiterhin gilt mein Dank Michael Königsmann und seinen Kollegen bei Keysight Technologies für die großartige Unterstützung bei unseren Messungen. Ebenso möchte ich mich bei Joachim Lutz und den Mitarbeitern bei Micram für ihre Unterstützung bedanken.

Meinen Kollegen am IPQ gebührt ein großer Dank. Die tolle Atmosphäre und Zusammenarbeit haben die Zeit am Institut nicht nur zu reiner Arbeit werden lassen. Vielen Dank für die Kickerspiele, das Grillen auf dem Dach, die Runden an der Kaffeemaschine und all die Ausflüge und sonstige Unternehmungen.

Vielen Dank an Robert Palmer, Dietmar Korn und Luca Alloatti, von ihnen konnte ich den Staffelstab der SOH Modulatoren übernehmen und weitertragen. Ohne eure Arbeit wäre die jetzige nicht möglich gewesen. Bei Stefan Wolf möchte ich mich für die Zusammenarbeit im Systemlabor und darüber hinaus bedanken. Auch wenn die Modulatoren noch nicht ganz ‚System' waren, konnte ich mir immer ein Eckchen im Labor greifen. Danke auch an Wladick Hartmann und Aleksandar Nesic für die gemeinsame Arbeit an den Maskenlayouts vor, während, und über die Deadlines hinaus. Dank Dr. Sebastian Köber und Dr. Yasar Kutuvantavida konnte ich viel über elektrooptische Materialien und deren Stabilität lernen, was ein wesentlicher Baustein für meine Arbeit war.

Vielen Dank an das HLB Team, Sascha Mühlbrandt, Wladick Hartmann, Robert Palmer und Simon Schneider. Gemeinsam haben wir es doch immer wieder geschafft, schöne Übungen und Klausuren zu basteln.

Vielen Dank an alle meine Mitdoktoranden am Institut über die Jahre, Oding Billah, Matthias Blaicher, Philipp Dietrich, Tobias Harter, Tobias Hoose, Juned Kemal, Clemens Kieninger, Kira Köhnle, Daria Kohler, Swen König, Nicole Lindenmann, Alexandra Ludwig, Pablo Marin, Argishti Melikyan, Jörg Pfeifle,

Philipp Schindler, Rene Schmogrow, Sandeep Ummethala, Claudius Weimann, Tobias Wienhold und Heiner Zwickel. Es war großartig die Zeit mit euch zu verbringen.

Ich danke meinen Studenten Niklas Baumgarten, Min Chen, Daniel Hilbert, Kristina Geistert und Christian Puterity für ihre Arbeit an den SOH Modulatoren, Alexander Knopf für die Arbeit am Frequenzschieber und Lu Zhou das Weiterverfolgen des SOH Lasers.

Herzlichst möchte ich mich bei Bernadette Lehmann, Andrea Riemensperger und Tatiana Gassmann, dem Sekretariatsteam des Instituts bedanken. Sie schafften es in allen Situationen den Überblick zu behalten und bei Problemen jeglicher Art schnell weiter zu helfen.

Vielen Dank an Martin Winkeler und David Guder die bei Fragen zur Elektronik oder EDV immer weiterhelfen und Probleme schnell lösen konnten. Danke an Oswald Speck, der beim Bearbeiten von Chips und Fasern Präzisionsarbeit leistete. Vielen Dank auch an Florian Rupp für die Arbeit im Reinraum, an der Wafersäge und am REM. Vielen Dank an das Team der Werkstatt um Marco Hummel und Manfred Hirsch. Ohne die gekonnten Arbeiten der Werkstatt wären die Labore sehr leer und viele Experimente nicht möglich.

Ein großer Dank gilt meiner Mutter Renate und meinem Vater Paul. Egal was ich vorhatte, gemacht habe oder auch nicht gemacht habe, ich konnte und kann mir immer sicher sein, dass ihr mich unterstützt und dass ihr hinter mir steht. Vielen Dank dafür.

Selina, vielen Dank für deine Liebe, dein Verständnis und deine Unterstützung während dieser Arbeit und darüber hinaus.

List of publications

Patents

[P1] J. Li, **M. Lauermann**, S. Schuele, J. Leuthold, and W. Freude, "Optical detector for detecting optical signal beams, method to detect optical signals, and use of an optical detector to detect optical signals," US patent 20, 120, 224, 184, (2012).

Book chapters

[B1] L. R. Dalton, **M. Lauermann**, and C. Koos. "NLO: Electro-Optic Applications" In: Bredas, J. -L.; Marder, S. R. (Eds): "The WSPC Reference on Organic Electronics: Organic Semiconductors", Volume 2: Fundamental Aspects of Materials and Applications, *World Scientific,* ISBN 978-981-4699-22-8, (2016).

Journal publications

[J1] J. Wang, M. Kroh, A. Theurer, C. Zawadzki, D. Schmidt, R. Ludwig, **M. Lauermann**, Z. Zhang, A. Beling, A. Matiss, C. Schubert, A. Steffan, N. Keil, and N. Grote, "Dual-quadrature coherent receiver for 100G Ethernet applications based on polymer planar lightwave circuit," *Optics Express*, vol. 19, no. 26, p. B166, Dec. 2011.

[J2] J. Leuthold, C. Koos, W. Freude, L. Alloatti, R. Palmer, D. Korn, J. Pfeifle, **M. Lauermann**, R. Dinu, S. Wehrli, M. Jazbinsek, P. Gunter, M. Waldow, T. Wahlbrink, J. Bolten, H. Kurz, M. Fournier, J.-M. Fedeli, H. Yu, and W. Bogaerts, "Silicon-organic hybrid electro-optical devices," *IEEE J. Sel. Topics Quantum Electron.*, vol. 19, no. 6, pp. 114–126, Nov. 2013.

[J3] L. Alloatti, **M. Lauermann**, C. Sürgers, C. Koos, W. Freude, and J. Leuthold, "Optical absorption in silicon layers in the presence of charge inversion/accumulation or ion implantation," *Appl. Phys. Lett.*, vol. 103, no. 5, p. 051104, 2013.

[J4] J. Li, M. R. Billah, P. C. Schindler, **M. Lauermann**, S. Schuele, S. Hengsbach, U. Hollenbach, J. Mohr, C. Koos, W. Freude, and J. Leuthold, "Four-in-one interferometer for coherent and self-coherent detection," *Optics Express*, vol. 21, no. 11, p. 13293, Jun. 2013.

[J5] J. Pfeifle, V. Brasch, **M. Lauermann**, Y. Yu, D. Wegner, T. Herr, K. Hartinger, P. Schindler, J. Li, D. Hillerkuss, R. Schmogrow, C. Weimann, R. Holzwarth, W. Freude, J. Leuthold, T. J. Kippenberg, and C. Koos, "Coherent terabit communications with microresonator Kerr frequency combs," *Nature Photonics*, vol. 8, no. 5, pp. 375–380, Apr. 2014.

[J6] R. Palmer, S. Koeber, D. Elder, M. Woessner, W. Heni, D. Korn, **M. Lauermann**, W. Bogaerts, L. Dalton, W. Freude, J. Leuthold, and C. Koos, "High-speed, low drive-voltage silicon-organic hybrid modulator based on a binary-chromophore electro-optic material," *J. Lightw. Technol.*, vol. 32, no. 16, pp. 2726–2734, 2014.

[J7] **M. Lauermann**, R. Palmer, S. Koeber, P. C. Schindler, D. Korn, T. Wahlbrink, J. Bolten, M. Waldow, D. L. Elder, L. R. Dalton, J. Leuthold, W. Freude, and C. Koos, "Low-power silicon-organic hybrid (SOH) modulators for advanced modulation formats," *Opt. Express*, vol. 22, no. 24, p. 29927, Dec. 2014.

[J8] S. Koeber, R. Palmer, **M. Lauermann**, W. Heni, D. L. Elder, D. Korn, M. Woessner, L. Alloatti, S. Koenig, P. C. Schindler, H. Yu, W. Bogaerts, L. R. Dalton, W. Freude, J. Leuthold, and C. Koos, "Femtojoule electro-optic modulation using a silicon–organic hybrid device," *Light Sci. Appl.*, vol. 4, no. 2, p. e255, Feb. 2015.

[J9] **M. Lauermann**, S. Wolf, P. C. Schindler, R. Palmer, S. Koeber, D. Korn, L. Alloatti, T. Wahlbrink, J. Bolten, M. Waldow, M. Koenigsmann, M. Kohler, D. Malsam, D. L. Elder, P. V. Johnston, N. Phillips-Sylvain, P. A. Sullivan, L. R. Dalton, J. Leuthold, W. Freude, and C. Koos, "40 GBd 16QAM signaling at 160 Gb/s in a silicon-organic hybrid modulator," *J. Lightw. Technol.*, vol. 33, no. 6, pp. 1210–1216, Mar. 2015.

[J10] S. Wolf, **M. Lauermann**, P. Schindler, G. Ronniger, K. Geistert, R. Palmer, S. Kober, W. Bogaerts, J. Leuthold, W. Freude, and C. Koos, "DAC-less amplifier-less generation and transmission of QAM signals using sub-volt silicon-organic hybrid modulators," *J. Lightw. Technol.*, vol. 33, no. 7, pp. 1425–1432, Apr. 2015.

[J11] A. Melikyan, K. Koehnle, **M. Lauermann**, R. Palmer, S. Koeber, S. Muehlbrandt, P. C. Schindler, D. L. Elder, S. Wolf, W. Heni, C. Haffner, Y. Fedoryshyn, D. Hillerkuss, M. Sommer, L. R. Dalton, D. Van Thourhout, W. Freude, M. Kohl, J. Leuthold, and C. Koos, "Plasmonic-organic hybrid (POH) modulators for OOK and BPSK signaling at 40 Gbit/s," *Opt. Express*, vol. 23, no. 8, p. 9938, Apr. 2015.

[J12] L. Alloatti, C. Kieninger, A. Froelich, **M. Lauermann**, T. Frenzel, K. Köhnle, W. Freude, J. Leuthold, M. Wegener, and C. Koos, "Second-order nonlinear optical metamaterials: ABC-type nanolaminates," *Applied Physics Letters*, vol. 107, no. 12, p. 121903, Sep. 2015.

[J13] C. Koos, J. Leuthold, W. Freude, M. Kohl, L. Dalton, W. Bogaerts, A. Giesecke, **M. Lauermann**, A. Melikyan, S. Koeber, S. Wolf, C. Weimann, S. Muehlbrandt, K. Koehnle, J. Pfeifle, W. Hartmann, Y. Kutuvantavida, S. Ummethala, R. Palmer, D. Korn, L. Alloatti, P. Schindler, D. Elder, T. Wahlbrink, and J. Bolten, "Silicon-organic hybrid (SOH) and plasmonic-organic hybrid (POH) integration," *J. Lightw. Technol.*, vol. 34, no. 2, pp. 256–268, Jan. 2016.

[J14] S. Schneider, **M. Lauermann**, P.-I. Dietrich, C. Weimann, W. Freude, and C. Koos, "Optical coherence tomography system mass-producible on a silicon photonic chip," *Optics Express*, vol. 24, no. 2, p. 1573, Jan. 2016.

[J15] D. Korn, **M. Lauermann**, S. Koeber, P. Appel, L. Alloatti, R. Palmer, P. Dumon, W. Freude, J. Leuthold, and C. Koos, "Lasing in silicon–organic hybrid waveguides," *Nature Communications*, vol. 7, p. 10864, Mar. 2016.

[J16] **M. Lauermann**, S. Wolf, W. Hartmann, R. Palmer, Y. Kutuvantavida, H. Zwickel, A. Bielik, L. Altenhain, J. Lutz, R. Schmid, T. Wahlbrink, J. Bolten, A. L. Giesecke, W. Freude, and C. Koos, "Generation of 64 GBd 4ASK signals using a silicon-organic hybrid modulator at 80°C," *Optics Express*, vol. 24, no. 9, p. 9389, May 2016.

[J17] **M. Lauermann**, C. Weimann, A. Knopf, W. Heni, R. Palmer, S. Koeber, D. L. Elder, W. Bogaerts, J. Leuthold, L. R. Dalton, C. Rembe, W. Freude, and C. Koos, "Integrated optical frequency shifter in silicon-organic hybrid (SOH) technology," *Optics Express*, vol. 24, no. 11, p. 11694, May 2016.

Conference publications

[C1] J. Li, C. Schmidt-Langhorst, R. Schmogrow, D. Hillerkuss, **M. Lauermann**, M. Winter, K. Worms, C. Schubert, C. Koos, W. Freude, and J. Leuthold, "Self-Coherent Receiver for PolMUX Coherent Signals," in Optical Fiber Communication Conference 2011, p. OWV5.

[C2] F. Soares, Z. Zhang, G. Przyrembel, **M Lauermann**, M. Moehrle, C. Zawadzki, B. Zittermann, N. Keil, and N. Grote, "Hybrid photonic integration of InP-based laser diodes and polymer PLCs," in Compound Semiconductor Week (CSW/IPRM) and 23rd International Conference on Indium Phosphide and Related Materials, 2011, pp. 1–4.

[C3] M. Kroh, J. Wang, A. Theurer, C. Zawadzki, D. Schmidt, R. Ludwig, **M. Lauermann**, A. Beling, A. Matiss, C. Schubert, A. Steffan, N. Keil, and N. Grote, "Coherent Receiver for 100G Ethernet Applications Based on Polymer Planar Lightwave Circuit," in 37th European Conference and Exposition on Optical Communications (ECOC'11) 2011, p. Tu.3.LeSaleve.2.

[C4] J. Wang, **M. Lauermann**, C. Zawadzki, W. Brinker, Z. Zhang, D. de Felipe, N. Keil, N. Grote, and M. Schell, "Quantifying direct DQPSK receiver with integrated photodiode array by assessing an adapted common-mode rejection ratio," in Communications and Photonics Conference and Exhibition, (ACP 2011) 2011, p. 83080R.

[C5] D. Korn, **M. Lauermann**, P. Appel, L. Alloatti, R. Palmer, W. Freude, J. Leuthold, and C. Koos, "First Silicon-Organic Hybrid Laser at Telecommunication Wavelength," in Conference on Lasers and Electro-Optics (CLEO'12) 2012, p. CTu2I.1.

[C6] **M. Lauermann**, D. Korn, P. Appel, L. Alloatti, W. Freude, J. Leuthold, and C. Koos, "Silicon-Organic Hybrid (SOH) Lasers at Telecommunication Wavelengths," in Conference on Integrated Photonics Research, Silicon and Nano-Photonics (IPR'12) 2012, p. IM3A.3.

[C7] D. Korn, L. Alloatti, **M. Lauermann**, J. Pfeifle, R. Palmer, P. C. Schindler, W. Freude, C. Koos, J. Leuthold, H. Yu, W. Bogaerts, K. Komorowska, R. Baets, J. Van Campenhout, P. Verheyen, J. Wouters, M. Moelants, P. Absil, A. Secchi, M. Dispenza, M. Jazbinsek, P. Gunter, S. Wehrli, M. Bossard, P. Zakynthinos, and I. Tomkos, "Silicon-organic hybrid fabrication platform for integrated circuits," in 14th International Conference on Transparent Optical Networks (ICTON) 2012, pp. 1–4.

[C8] C. Koos, J. Leuthold, W. Freude, L. Alloatti, D. Korn, R. Palmer, **M. Lauermann**, N. Lindenmann, S. Koeber, J. Pfeifle, P. C. Schindler, D. Hillerkuss, and R. Schmogrow, "Silicon-Organic Hybrid Integration and Photonic Wire Bonding: Technologies for Terabit/s Interconnects," in Joint Symposium on Opto- and Microelectronic Dvices and Circuits (SODC2012), Hangzhou, China, 2012.

[C9] J. Pfeifle, **M. Lauermann**, D. Wegner, J. Li, K. Hartinger, V. Brasch, T. Herr, D. Hillerkuss, R. M. Schmogrow, T. Schimmel, R. Holzwarth, T. J. Kippenberg, J. Leuthold, W. Freude, and C. Koos, "Microresonator-Based Frequency Comb Generator as Optical Source for Coherent WDM Transmission," in Optical Fiber Communication Conference (OFC'13) 2013, p. OW3C.2.

[C10] J. Leuthold, C. Koos, W. Freude, L. Alloatti, R. Palmer, D. Korn, J. Pfeifle, **M. Lauermann**, R. Dinu, S. Wehrli, M. Jazbinsek, P. Gunter, M. Waldow, T. Wahlbrink, J. Bolten, M. Fournier, J. M. Fedeli, W. Bogaerts, and H. Yu, "High-speed, low-power optical modulators in silicon," in 15th International Conference on Transparent Optical Networks (ICTON) 2013, pp. 1–4.

[C11] C. Koos, J. Leuthold, W. Freude, L. Alloatti, R. Palmer, D. Korn, J. Pfeifle, P. C. Schindler, and **M. Lauermann**, "Silicon-organic hybrid (SOH) technology: A platform for efficient electro-optical devices," in International Conference on Optical MEMS and Nanophotonics (OMN) 2013, pp. 85–86.

[C12] W. Freude, L. Alloatti, D. Korn, **M. Lauermann**, A. Melikyan, R. Palmer, J. Pfeifle, P. C. Schindler, C. Weimann, R. Dinu, J. Bolten, T. Wahlbrink, M. Waldow, S. Walheim, P. Leufke, S. Ulrich, J. Ye, P. Vincze, H. Hahn, H. Yu, W. Bogaerts, V. Brasch, T. Herr, R. Holzwarth, K. Hartinger, C. Stamatiadis, M. O'Keefe, L. Stampoulidis, L. Zimmermann, R. Baets, T. Schimmel, I. Tomkos, K. Petermann, T. J. Kippenberg, C. G. Koos, and J. Leuthold, "Nonlinear Nano-Photonics," in Frontiers in Optics (FiO'13) 2013, p. FTu4C.6.

[C13] C. G. Koos, W. Freude, J. Leuthold, L. Alloatti, R. Palmer, D. Korn, J. Pfeifle, P. C. Schindler, **M. Lauermann**, N. Lindenmann, S. Koeber, T. Hoose, M. R. Billah, H. Yu, W. Bogaerts, R. Baets, M. Fournier, J.-M. Fedeli, R. Dinu, J. Bolten, T. Wahlbrink, and M. Waldow, "Silicon-organic hybrid integration and photonic wire bonding: Enabling technologies for heterogeneous photonic systems," in Frontiers in Optics (FiO'13) 2013, p. FTu1E.3.

[C14] **M. Lauermann**, C. Weimann, A. Knopf, D. L. Elder, W. Heni, R. Palmer, D. Korn, P. Schindler, S. Koeber, L. Alloatti, H. Yu, W. Bogaerts, L. R. Dalton, C. Rembe, J. Leuthold, W. Freude, and C. Koos, "Integrated Silicon-Organic Hybrid (SOH) Frequency Shifter," in Optical Fiber Communication Conference (OFC'14) 2014, p. Tu2A.1.

[C15] R. Palmer, S. Koeber, M. Woessner, D. L. Elder, W. Heni, D. Korn, H. Yu, **M. Lauermann**, W. Bogaerts, L. R. Dalton, W. Freude, J. Leuthold, and C. Koos, "High-Speed Silicon-Organic Hybrid (SOH) Modulators with 230 pm/V Electro-Optic Coefficient Using Advanced Materials," in Optical Fiber Communication Conference (OFC'14) 2014, p. M3G.4.

[C16] **M. Lauermann**, R. Palmer, S. Koeber, P. C. Schindler, D. Korn, T. Wahlbrink, J. Bolten, M. Waldow, D. L. Elder, L. R. Dalton, J. Leuthold, W. Freude, and C. G. Koos, "16QAM silicon-organic hybrid (SOH) modulator operating with 0.6 Vpp and 19 fJ/bit at 112 Gbit/s," in Conf. on Lasers and Electro-Optics (CLEO'14), 2014, p. SM2G.6.

[C17] C. Weimann, **M. Lauermann**, T. Fehrenbach, R. Palmer, F. Hoeller, W. Freude, and C. G. Koos, "Silicon Photonic Integrated Circuit for Fast Distance Measurement with Frequency Combs," in Conf. on Lasers and Electro-Optics (CLEO'14), 2014, p. STh4O.3.

[C18] S. Schneider, **M. Lauermann**, C. Weimann, W. Freude, and C. G. Koos, "Silicon Photonic Optical Coherence Tomography System," in Conf. on Lasers and Electro-Optics (CLEO'14),, 2014, p. ATu2P.4.

[C19] **M. Lauermann**, C. Weimann, R. Palmer, P. C. Schindler, S. Koeber, C. Rembe, W. Freude, and C. Koos, "Integrated nanophotonic frequency shifter on the silicon-organic hybrid (SOH) platform for laser vibrometry," in 11th International Conference on Vibration Measurements by Laser and Noncontact Techniques (AIVELA'14) 2014, pp. 426–430.

[C20] S. Wolf, P. C. Schindler, G. Ronniger, **M. Lauermann**, R. Palmer, S. Koeber, D. Korn, W. Bogaerts, J. Leuthold, W. Freude, and C. Koos, "10 GBd SOH modulator directly driven by an FPGA without electrical amplification," in 40th European Conference on Optical Communication (ECOC 2014), 2014, pp. 1–3.

[C21] **M. Lauermann**, P. C. Schindler, S. Wolf, R. Palmer, S. Koeber, D. Korn, L. Alloatti, T. Wahlbrink, J. Bolten, M. Waldow, M. Koenigsmann, M. Kohler, D. Malsam, D. L. Elder, P. V. Johnston, N. Phillips-Sylvain, P. A. Sullivan, L. R. Dalton, J. Leuthold, W. Freude, and C. Koos, "40 GBd 16QAM modulation at 160 Gbit/s in a silicon-organic hybrid (SOH) modulator," in 40th European Conference on Optical Communication (ECOC 2014), France, 2014, p. We.3.1.3.

[C22] S. Wolf, **M. Lauermann**, G. Ronniger, P. C. Schindler, R. Palmer, S. Koeber, W. Bogaerts, J. Leuthold, W. Freude, and C. Koos, "DAC-less and amplifier-less generation and transmission of 16QAM signals using a sub-volt silicon photonic modulator," in 40th European Conference on Optical Communication (ECOC 2014), 2014, pp. 1–3.

[C23] J. Leuthold, A. Melikyan, L. Alloatti, D. Korn, R. Palmer, D. Hillerkuss, **M. Lauermann**, P. C. Schindler, B. Chen, R. Dinu, D. L. Elder, L. R. Dalton, C. Koos, M. Kohl, W. Freude, and C. Hafner, "From silicon-organic hybrid to plasmonic modulation," in 40th European Conference on Optical Communication (ECOC 2014), 2014, pp. 1–3.

[C24] P. C. Schindler, **M. Lauermann**, S. Wolf, D. Korn, R. Palmer, S. Koeber, W. Heni, A. Ludwig, R. Schmogrow, D. L. Elder, L. R. Dalton, W. Bogaerts, H. Yu, W. Freude, J. Leuthold, and C. Koos, "Ultra-short silicon-organic hybrid (SOH) modulator for bidirectional polarization-independent operation," in 40th European Conference on Optical Communication (ECOC 2014), 2014, pp. 1–3.

[C25] C. Koos, W. Freude, T. J. Kippenberg, J. Leuthold, L. R. Dalton, J. Pfeifle, C. Weimann, **M. Lauermann**, R. Palmer, S. Koeber, S. Wolf, P. Schindler, V. Brasch, and D. Elder, "Terabit/s optical transmission using chip-scale frequency comb sources," in 40th European Conference on Optical Communication (ECOC 2014), 2014, pp. 1–3.

[C26] K. Koehnle, **M. Lauermann**, A. Bacher, M.-K. Nees, S. Muehlbrandt, R. Palmer, A. Muslija, P.J. Jakobs, W. Freude, and C. Koos, "Silicon-Organic Hybrid Phase Modulator Based on Sub-Wavelength Grating Waveguides," in Micro and Nano Engineering (MNE2014), 2014, p. 8127 D2L–B.

[C27] C. Koos, T. J. Kippenberg, L. P. Barry, L. Dalton, W. Freude, J. Leuthold, J. Pfeifle, C. Weimann, **M. Lauermann**, J. N. Kemal, R. Palmer, S. Koeber, P. C. Schindler, T. Herr, V. Brasch, R. T. Watts, and D. Elder, "Terabit/s communications using chip-scale frequency comb sources," in Laser Resonators, Microresonators, and Beam Control XVII (LASE-SPIE'15), 2015, p. 93430E.

[C28] **M. Lauermann**, S. Wolf, R. Palmer, A. Bielik, L. Altenhain, J. Lutz, R. Schmid, T. Wahlbrink, J. Bolten, A. L. Giesecke, W. Freude, and C. Koos, "64 GBd operation of a silicon-organic hybrid modulator at elevated temperature," in Optical Fiber Communication Conference (OFC2015), 2015, p. Tu2A.5.

[C29] C. Koos, J. Leuthold, W. Freude, M. Kohl, L. Dalton, W. Bogaerts, A.-L. Giesecke, **M. Lauermann**, A. Melikyan, S. Koeber, S. Wolf, C. Weimann, S. Muehlbrandt, K. Koehnle, J. Pfeifle, R. Palmer, L. Alloatti, D. Elder, T. Wahlbrink, and J. Bolten, "Silicon-Organic Hybrid (SOH) and Plasmonic-Organic Hybrid (POH) Integration," in Optical Fiber Communication Conference (OFC2015), 2015, p. Tu2A.1.

[C30] **M. Lauermann**, S. Wolf, R. Palmer, S. Koeber, P. C. Schindler, T. Wahlbrink, J. Bolten, A. L. Giesecke, M. Koenigsmann, M. Kohler, D. Malsam, D. L. Elder, L. R. Dalton, J. Leuthold, W. Freude, and C. Koos, "High-speed and low-power silicon-organic hybrid modulators for advanced modulation formats," in SPIE 9516 Integrated Optics: Physics and Simulations II, 2015, p. 951607.

[C31] A. Melikyan, K. Köhnle, **M. Lauermann**, R. Palmer, S. R. Koeber, S. Muehlbrandt, P. Schindler, D. Elder, S. Wolf, W. Heni, C. Haffner, Y. Fedoryshyn, D. Hillerkuss, M. Sommer, L. Dalton, D. VanThourhout, W. Freude, M. Kohl, J. Leuthold, and C. Koos, "Plasmonic-organic hybrid (POH) modulators for OOK and BPSK signaling at 40 Gbit/s," in Conf. on Lasers and Electro-Optics (CLEO'15) 2015, p. SM1I.1.

[C32] C. Koos, W. Freude, J. Leuthold, M. Kohl, L. R. Dalton, W. Bogaerts, **M. Lauermann**, S. Wolf, R. Palmer, S. Koeber, A. Melikyan, C. Weimann, G. Ronniger, K. Geistert, P. C. Schindler, D. L. Elder, T. Wahlbrink, J. Bolten, A. L. Giesecke, M. Koenigsmann, M. Kohler, and D. Malsam, "Silicon-organic hybrid (SOH) integration for low-power and high-speed signal generation," in 17th Intern. Conf. on Transparent Optical Networks (ICTON'15), 2015, pp. 1–2.

[C33] C. Koos, T. Kippenberg, L. P. barry, L. Dalton, A. Ramdane, F. Lelarge, J. Leuthold, W. Freude, J. Pfeifle, C. Weimann, J. N. Kemal, **M. Lauermann**, S. Wolf, I. Shkarban, S. Koeber, T. Herr, V. Brasch, R. Watts, and D. Elder, "Coherent Terabit Communications using Chip-scale Frequency Comb Sources," in OSA Topical Meeting on Nonlinear Optics (NLO'15), 2015, p. NTh2A.1.

[C34] W. Hartmann, **M. Lauermann**, S. Wolf, H. Zwickel, Y. Kutuvantavida, J. Luo, A. K.-Y. Jen, W. Freude, and C. Koos, "100 Gbit/s OOK using a silicon-organic hybrid (SOH) modulator," in European Conference on Optical Communication (ECOC), 2015, Valencia, Spain, 2015, PDP.1.4.

[C35] C. Koos, W. Freude, J. Leuthold, M. Kohl, L. R. Dalton, W. Bogaerts, **M. Lauermann**, S. Wolf, C. Weimann, A. Melikyan, N. Lindenmann, M. R. Billah, S. Muehlbrandt, S. Koeber, R. Palmer, K. Koehnle, L. Alloatti, D. L. Elder, A.-L. Giesecke, and T. Wahlbrink, "Silicon-organic hybrid (SOH) integration and photonic multi-chip systems: Extending the capabilities of the silicon photonic platform," in 28th Annual Conf. of the IEEE Photonics Society (IPC'15), 2015, pp. 480–481.

[C36] L. Alloatti, C. M. Kieninger, A. M. Frölich, **M. Lauermann**, T. Frenzel, K. Köhnle, W. Freude, J. Leuthold, C. Koos, and M. Wegener, "Second-harmonic generation from atomic-scale ABC-type laminate optical metamaterials," in SPIE 9544, Metamaterials, Metadevices, and Metasystems 2015, 2015, p. 95440Y.

[C37] S. Wolf, **M. Lauermann**, W. Hartmann, H. Zwickel, Y. Kutuvantavida, M. Koenigsmann, M. Gruen, J. Luo, A. K. Jen, W. Freude, and C. Koos, "An energy-efficient 252 Gbit/s silicon-based IQ-modulator," in Optical Fiber Communication Conference (OFC) 2016, USA, 2016, p. Th3J.2.

[C38] C. Koos, W. Freude, J. Leuthold, L. R. Dalton, S. Wolf, H. Zwickel, T. Hoose, M. R. Billah, **M. Lauermann**, C. Weimann, W. Hartmann, A. Melikyan, N. Lindenmann, S. Koeber, R. Palmer, L. Alloatti, A. L. Giesecke, T. Wahlbrink, "Silicon-organic hybrid (SOH) integration and photonic multi-chip systems: Technologies for high-speed optical interconnects," in Optical Interconnects Conference (OI'16), Hyatt Regency Mission Bay Spa & Marina, San Diego, California, USA, (2016) (invited)

[C39] W. Freude, S. Schneider, **M. Lauermann**, P.-I. Dietrich, C. Weimann, C. Koos, „Silicon photonic integrated circuit for optical coherence tomography", 18th Intern. Conf. on Transparent Optical Networks (ICTON'16), Trento, Italy, paper Tu.C5.1 (2016) (invited)

[C40] C. Koos, T. J. Kippenberg, L. P. Barry, L. R. Dalton, A. Ramdane, F. Lelarge, W. Freude, J. N. Kemal, P. Marin, S. Wolf, P. Trocha, J. Pfeifle, C. Weimann, **M. Lauermann**, T. Herr, V. Brasch, R. T. Watts, D. L. Elder, A. Martinez, V. Panapakkam, N. Chimot, "Multi-terabit/s transmission using chip-scale frequency comb sources," in 18th Intern.

Conf. on Transparent Optical Networks (ICTON'16), Trento, Italy, paper We.A5.1 (2016) (invited)

[C41] A. Nesic, M. Blaicher, T. Hoose, **M. Lauermann**, Y. Kutuvantavida, W. Freude, C. Koos, „Hybrid 2D/3D photonic integration for non-planar circuit topologies" in 42th European Conf. Opt. Commun. (ECOC'16), Düsseldorf, Germany, paper W.3.F.4 (2016)

[C42] H. Zwickel, S. Wolf, Y. Kutuvantavida, C. Kieninger, **M. Lauermann**, W. Freude; C. Koos, "120 Gbit/s PAM-4 signaling using a silicon-organic hybrid (SOH) Mach-Zehnder modulator" in 42th European Conf. Opt. Commun. (ECOC'16), Düsseldorf, Germany, paper Th2.P2.SC2.14 (2016)

[C43] S. Wolf, W. Hartmann, **M. Lauermann**, H. Zwickel, Y. Kutuvantavida, C. Kieninger, W. Freude, C. Koos, „High-speed silicon-organic hybrid (SOH) modulators" in 42th European Conf. Opt. Commun. (ECOC'16), Düsseldorf, Germany, paper Tu.3.A.3 (2016) (invited)

Curriculum vitae

Matthias Lauermann

born in Schwäbisch Hall, Germany

Nationality: German

m.lauermann@posteo.de

Education

03/2012–11/2015	Ph.D. candidate and research associate at the Institute of Photonics and Quantum Electronics at the Karlsruhe Institute of Technology. Field of study: Silicon Photonics
10/2005–12/2011	Studies at the Karlsruhe Institute of Technology, Germany Major: Electrical Engineering and Information Technology Specialization: optical communication technology Thesis: "Silicon organic hybrid laser" Degree: Dipl.-Ing.
08/2008–12/2008	Studies at Royal Institute of Technology in Stockholm

School education

09/1995–07/2004	Hohenlohe Gymnasium Öhringen, Germany Degree: Abitur (general university entrance qualification)

Honors and awards

Maiman Outstanding Student Paper Competition:
Honorable mention at Conference on Lasers and Electro-Optics (CLEO) 2014

CPSIA information can be obtained
at www.ICGtesting.com
Printed in the USA
LVHW022001170821
695476LV00007B/117